The Vanishing Face of Gaia

JAMES LOVELOCK

The Vanishing Face of Gaia

A Final Warning

with a Foreword by Martin Rees

ALLEN LANE
an imprint of
PENGUIN BOOKS

ALLEN LANE

Published by the Penguin Group
Penguin Books Ltd, 80 Strand, London WC2R ORL, England
Penguin Group (USA) Inc., 375 Hudson Street, New York, New York 10014, USA
Penguin Group (Canada), 90 Eglinton Avenue East, Suite 700, Toronto, Ontario, Canada M4P 2Y3
(a division of Pearson Canada Inc.)
Penguin Ireland, 25 St Stephen's Green, Dublin 2, Ireland (a division of Penguin Books Ltd)
Penguin Group (Australia), 250 Camberwell Road, Camberwell, Victoria 3124, Australia
(a division of Pearson Australia Group Pty Ltd)
Penguin Books India Pvt Ltd, 11 Community Centre, Panchsheel Park, New Delhi – 110 017, India
Penguin Group (NZ), 67 Apollo Drive, Rosedale, North Shore 0632, New Zealand
(a division of Pearson New Zealand Ltd)
Penguin Books (South Africa) (Pty) Ltd, 24 Sturdee Avenue,
Rosebank, Johannesburg 2196, South Africa

Penguin Books Ltd, Registered Offices: 80 Strand, London WC2R ORL, England

www.penguin.com

First published 2009
4

Copyright © James Lovelock, 2009
Foreword copyright © Martin Rees, 2009

The moral right of the author has been asserted

Set in 10.5/14 pt Linotype Sabon
Typeset by Rowland Phototypesetting Ltd, Bury St Edmunds, Suffolk
Printed in England by Clays Ltd, St Ives plc

HARDBACK ISBN: 978-1-846-14185-0
TRADE PAPERBACK ISBN: 978-1-846-14227-7

www.greenpenguin.co.uk

Penguin Books is committed to a sustainable future
for our business, our readers and our planet.
The book in your hands is made from paper
certified by the Forest Stewardship Council.

For my beloved wife Sandy

Contents

List of Illustrations   ix
Acknowledgements   x
Foreword by Martin Rees   xi

1 The Journey in Space and Time   1
2 The Climate Forecast   23
3 Consequences and Survival   46
4 Energy and Food Sources   64
5 Geoengineering   92
6 The History of Gaia Theory   105
7 Perceptions of Gaia   123
8 To Be or Not To Be Green   134
9 To the Next World   149

Glossary   163
Further Reading   169
Index   173

List of Illustrations

1. The 'Ice Hole' in the Arctic. (US National Snow and Ice Data Centre)
2. The vanishing of the Aral Sea. (NASA)
3. White Knight 2.
4. The Virgin Galactic space ship.
5. The Devon countryside.
6. A wind farm in Scotland.
7. The electron capture detector, or ECD.
8. The author standing above French high-level nuclear waste.
9. Drought in Australia. (Wikipedia. GNU Free Documentation Licence)
10. The Namibian desert.
11. Vegetation types around the world. (Wikipedia. GNU Free Documentation Licence)
12. The world's oceans showing vast areas of barren water.
13. Vegetation in the wild in central Brazil. (Wikipedia. GNU Free Documentation Licence)

Acknowledgements

I am truly grateful to Richard Betts, John Gray, Armand Neukermans, Sir Crispin Tickell, Brian Foulger, Gari Owen, Tim Donaldson and Elaine Steel, who read the book and made helpful comments, and to Chris Rapley, Stephan Harding, Peter Liss, Andrew Watson, Tim Lenton and Dave Wilkinson for their valued advice. I also thank GAIA, registered charity no. 327903, for support during the writing of this book.

Foreword
by Martin Rees

A little more than forty years ago, the Apollo 8 astronauts, while orbiting the Moon, photographed the whole Earth – its biosphere contrasting with the sterile moonscape where the astronauts left their footprints. The Apollo images raised global awareness that 'Spaceship Earth' was vulnerable, and that sustaining it was an ecological imperative. But there was a second important influence with similar global resonance – not an image, but an arresting and romantically titled new concept. This was Gaia – the idea that the Earth's biosphere behaves as though it were a single organism.

Gaia was the insight of a man who is undoubtedly one of the most original and influential living scientists: James Lovelock. He believes that our species is now putting the Earth under unprecedented stress, and that climate change could lead to a world with much impoverished ecology that is barely habitable by humans. More scarily (and more controversially) he claims that the 'point of no return' may already have been passed.

Our planet is nearly 4.5 billion years old. If some aliens had been watching it from afar ever since its birth, what would they have seen? Over nearly all that immense time changes were incessant, but generally gradual. The continents drifted; the ice cover waxed and waned; global temperatures rose and fell; species emerged, evolved and became extinct.

But in just a tiny sliver of the Earth's history – the last one millionth part, a few thousand years – the patterns of vegetation altered much faster than before. This signalled the start of agriculture. The pace of change accelerated as human populations rose and engaged in urban and industrial activity. Consumption of fossil fuels caused an anomalously fast build-up of carbon dioxide in the atmosphere; the climate changed, and the world started to heat up.

If they understood astrophysics, the aliens watching our planet could confidently predict that the biosphere would face doom when the Sun brightens, and eventually flares up into a 'red giant' star. But could they have forseen this unprecedented sudden 'fever' less than half way through the Earth's life – these human-induced changes, seemingly occurring with runaway speed?

And what might these hypothetical aliens witness in the next hundred years? Will spasms be followed by stability? If so, will our Earth settle into a state that still offers a habitat for humans? Or have our unplanned interventions irrevocably tipped the planet into a new and far hotter climatic state? If so, how many present species of amimals and plants will survive?

These issues – climate change and loss of biodiversity – have risen high on the international agenda. James Lovelock is helping to keep them there. He is a hero to many scientists – certainly to me. His individualistic career is a welcome counterpoint to the specialized, quasi-industrial style in which most research is conducted. In the 1960s he designed an instrument that was so sensitive at detecting minute traces of atmospheric pollutants that many colleagues refused to believe his claims. He is beholden to no institution. He ranges freely across the disciplinary boundaries that too often constrain 'institutional' thinkers.

The flavour of James Lovelock's mind and personality shine through in this important and highly readable book. He writes clearly – even entertainingly – with many apt analogies. But he writes also with passion; and his thoughts are grounded in a lifetime of distinguished work. He is both a fine scientist and an eloquent advocate of action.

Many of us still hope that our civilization can segue towards a low-carbon future and a lower population – and achieve this transition without trauma and disaster. But that benign outcome demands determined action by governments, urgently implemented; and such urgency won't be achieved unless sustained campaigning can transform public attitudes and lifestyles. Programmes to develop 'clean energy' must be accorded, worldwide, the urgency that the US gave to the Apollo programme in the 1960s.

Those of us who are scientists should aspire to emulate James Lovelock's inventiveness; all citizens should be inspired by his commitment and altruism. It is no exaggeration to say that our civilization's long-term future depends on whether the 'call to arms' in this riveting book is widely heeded.

Martin Rees
Trinity College, Cambridge
January 2009

I

The Journey in Space and Time

Icons are important to us: the cross and the scimitar have dominated lives and history for two millennia. For some the icon of greatest meaning is that blue-white vision of the Earth first seen from space by astronauts. That icon is undergoing subtle change as the white ice fades away, the green of forest and grassland fades into the dun of desert, and the oceans lose their blue-green hue and turn a purer, swimming-pool blue as they too become desert. This is why at ninety years old I will try to emulate the astronauts and fly into space to see the Earth from above before it vanishes. I want to catch a glimpse of the Earth I have lived with all my life, even though my most trusted physician, Douglas Chamberlain, has advised me that the risk is too high. I will go, despite the warnings, to recapture that enthralling eureka moment forty-four years ago when I was working at the then centre of space research, the Jet Propulsion Laboratory (JPL) in California and saw with my mind's eye our planet as something possibly unique in the universe, something alive. Since then I have thought of the word Earth as inadequate to describe the living planet that we inhabit and are a part of. I am grateful to the author William Golding for his suggestion that the name Gaia was more appropriate. Not least among the joys of seeing our live planet from high above will be the simple pleasure of seeing with my own eyes how spherical it is. I had few doubts that this was so but as with many things in life as well as in science, we have to take for granted that it is round, even though when on the ground, our eyes tell us it is flat.

Imagine my delighted amazement on hearing that my wish to see the Earth from space would soon be fulfilled and I would see from the sky above New Mexico our sphere of a world in all its glory. In an act of splendid generosity, Sir Richard Branson made the gift and had already founded his own space line, Virgin Galactic, to make it possible. His ultimate upgrade, the flight to space, will let me escape for a few brief minutes the all-pervasive introspection of 21st-century life and allow me to share that transcendental feeling of astronauts that our home is not the house or the street or the nation where we live, but the Earth itself.

Is there any need to see Gaia, the only living planet in the solar system? After all, despite the recent economic setback, life continues to get better in most of the world; even the poor in the developed world, although malnourished, are sometimes well enough fed to be obese. There are so many possibilities of entertainment that there is no reason to be bored day or night. Perhaps we no longer need to see the Earth in reality when we can see it so well on Google.

It does matter, and more than any other thing: we have to see it as it really is because our lives are wholly dependent upon the living Earth. We could not survive for an instant on a dead planet like Mars, and we need to understand the difference. If we fail to take our planet seriously we will be like children who take their homes for granted and never doubt that breakfast starts the day; we will not notice as we enjoy our daily lives that the cost of our neglect could soon cause the greatest tragedy in the memory of humankind. The Earth, in its but not our interests, may be forced to move to a hot epoch, one where it can survive, although in a diminished and less habitable state. If, as is likely, this happens, we will have been the cause.

Do not be misled by lulls in climate change when global temperature is constant for a few years or even, as I write here in the UK in 2008, appears to drop. Holidaymakers and farmers who have endured a miserably cloudy, cool, wet July and August all ask me: Where is global warming now? Further away in the Gulf of Mexico, where for several years the surface water was unusually warm, it is now cooler again, and the Arctic has also regained a little of the astounding losses of 2007 (although ominously the ice continues to grow thinner). In the real world change is rarely smooth: it goes by fits and starts, more like the halting progress of a traffic jam than the easy motion of the open road. But however unlikely it sometimes seems, change really is happening and the Earth grows warmer year by year. It is ever more at risk of changing to a barren state in which few of us can survive. Scientists, especially Steve Schneider and Jim Hansen, recognized in the 1980s the possibility of dangerous climate change as a result of our pollution of the air with excessive carbon dioxide. This led the eminent Swedish climatologist Bert Bolin to persuade the United Nations (UN) to form the Intergovernmental Panel on Climate Change (IPCC) with Sir John Houghton and Gylvan Meiro Filho as its first co-chairs. It began gathering evidence about the changing chemistry and physics of the atmosphere in 1990 and has issued reports in 1991, 1995, 2001 and 2007. Through the efforts of this more than 1,000-strong panel of scientists of many different nations we now know enough about the Earth's atmosphere to make intelligent guesses about future climates. But so far these guesses have been unable to match the observed changes in climate closely enough for us to be confident about IPCC forecasts decades into the future.

We are almost all of us now so urbanized that few living in the cities of the northern hemisphere see the stars at night. Air and light pollution have dimmed them so that only the Moon and Venus are visible through the night-time glare. Our great-grandparents often saw the constellations of stars and used Polaris to guide their way; on clear nights they even saw the Milky Way, that faint white band that crosses the sky and is a sideways view of our home galaxy. Apart from a few sailors and farmers miles away from any settlement, who still see the dark depths of the sky, we are all lost in the hazy air of that mega-city that globalization has made of the human world. In a similar way scientists have become urbanized and have only recently taken the idea of a live Earth into their thinking. Most of them have still to digest the idea of Gaia and make it part of their practice.

We are trying to undo some of the harm we have done, and as climate change worsens we will try harder, even desperately, but until we see that the Earth is more than a mere ball of rock we are unlikely to succeed. It is not simply too much carbon dioxide in the air or the loss of biodiversity as forests are cleared; the root cause is too many people, their pets and their livestock – more than the Earth can carry. No voluntary human act can reduce our numbers fast enough even to slow climate change. Merely by existing, people and their dependent animals are responsible for more than ten times the greenhouse gas emissions of all the airline travel in the world.

We do not seem to have the slightest understanding of the seriousness of our plight. Instead, before our thoughts were diverted by the global financial collapse, we seemed lost in an endless round of celebration and congratulation. It was good to recognize the huge efforts of the Intergovernmental Panel on Climate Change and Al Gore with the Nobel Peace Prize and to have a brave 10,000 make the long journey to Bali as a salutation, but because they failed to see the Earth as alive and responsive they ignored at our peril the extent of its disapproval of all we do. As we hold our meetings and talk of stewardship, Gaia still moves step by step towards the hot state, one that will allow her to continue as the regulator, but where few of us will be alive to meet and talk. Perhaps we were celebrating because the once rather worrying voice of the IPCC now spoke comfortably of consensus and endorsed those mysterious concepts of sustainability and energy that renewed itself. We even thought that this way somehow we could save the planet and grow richer as well, a more pleasing outcome than the uncomfortable truth.

I am not a willing Cassandra and in the past have been publicly sceptical about doom stories but this time we do have to take seriously the possibility that global heating may all but eliminate people from the Earth. It may seem that my pessimism is an extrapolation too far. I accept this: a continuing series of volcanic eruptions as powerful as Pinatubo in 1991 could reverse climate change, as might one or more of the geoengineering schemes now being considered; and possibly our projections are flawed. But pessimism is justified by the difference between the forecasts of the IPCC and what observers find in the real world. Just think, over 1,000 of the world's best climate scientists have worked for seventeen years to forecast future climates and have failed to predict the climate of today as I write in August 2008. I have little confidence in the smooth rising curve of temperature that modellers predict for the next ninety years. The Earth's history and simple climate models based on the notion of a live and responsive Earth suggest that sudden change and surprise are more likely. My pessimism is shared by other scientists and openly by the distinguished climate scientist James Hansen, who finds as I do that the evidence now coming from the Earth, together with the knowledge of its history, is gravely disturbing. Most of all I am pessimistic because business and governments both appear to be accepting uncritically a belief that climate change is easily and profitably reversible.

Do not expect the climate to follow the smooth path of slowly but sedately rising temperatures predicted by the IPCC, where change slowly inches up and leaves plenty of time for business as usual. The real Earth changes intermittently with spells of constancy, even slight decline, between the jumps to greater heat. Climate change is not at all like the smooth civil engineering of a major highway that climbs uninterruptedly up a mountain pass but more like the mountain itself, a concatenation of slopes, valleys, flat meadows, rock steps and precipices. Perhaps some time in the past the asset manager who cared for your pension fund showed you a growth curve for your investments that rose smoothly without a break from now until 2050; but now you would be full of doubt about so smooth and continuous a progress and would know that growth can be interrupted by Northern Rocks and Lehman Brothers strewn along the way and even fall into the chasm of a global recession. Yet we are asked to believe that temperature will rise smoothly for another forty years, unless of course we put the carbon dioxide in the atmosphere somewhere else. You may think that climate and economic forecasts have little in common, but they do: both systems are complex and non-linear and can change suddenly and unexpectedly. Alan Greenspan, until recently the USA's economic guru, said in a BBC interview that for this reason he refused to predict the course of the world economy; and the distinguished Cambridge economist Sir Partha Dasgupta warned that models of the economy shared with those of the climate a similar fickle unpredictability. They made these wise disclaimers well before the crash of 2008. We now know that the huge debts incurred by the first world were its cause. We have no notion when our environmental indebtedness will bring even greater ruin, only that it is likely to happen.

We seem to have forgotten that science is not wholly based on theory and models: more tiresome and prosaic confirmation by experiment and observation plays just as important a part. Perhaps for social reasons, science has in recent years changed its way of working. Observation in the real world and small-scale experiments on the Earth now take second place to expensive and ever-expanding theoretical models. It may be administratively and politically convenient to work this way but the consequences could be disastrous. Our tank is near empty of data and we are running on theoretical vapour: this is especially true of data about the oceans that make up over 70 per cent of the Earth's surface, and about the responses of ecosystems to climate change – and, just as importantly, the effect of change in the oceans and ecosystems on the climate.

The ideas that stem from Gaia theory put us in our proper place as part of the Earth system – not the owners, managers, commissars or people in charge. The Earth has not evolved solely for our benefit and any changes we make to it are at our own risk. This way of thinking makes clear that we have no special human rights; we are merely one of the partner species in the great enterprise of Gaia. We are creatures of Darwinian evolution, a transient species with a limited lifespan, as were all our numerous distant ancestors. But, unlike almost everything before we emerged on the planet, we are also intelligent social animals with the possibility of evolving to become a wiser and more intelligent animal, one that might have a greater potential as a partner for the rest of life on Earth. Our goal now is to survive and to live in a way that gives evolution beyond us the best chance. The philosopher John Gray has discussed the extent to which we are still an emerging intelligence and still have far to go to match even our own estimation of ourselves. Do we really believe that we humans, wholly untrained as we are, have the intelligence or capacity to manage the Earth?

We have become good at burying bad news and maybe this is why we do not like the reports brought back by those brave and true scientists who go out into the world, like Charles David Keeling and his son Ralph, who for so long and so accurately monitored carbon dioxide on the peak of Mauna Loa, or Andrew Watson taking wintertime measurements from a ship bouncing on the cold and stormy seas off Greenland. There are a few scientists like them who now make observations of temperature and sea level rise and their measurements were reported by Stefan Rahmstorf and his colleagues in May 2007 in *Science*. They found the sea level was rising 1.6 times as fast and the temperature 1.3 times as fast as the IPCC had predicted in 2007. In September 2007 we were devastated to discover that all but 40 per cent of the ice floating on the Arctic Ocean had melted. It is true that the visible loss in 2008 was slightly less, but the remaining ice had thinned by a record 1.5 feet. These changes are far more rapid than the gloomiest of model forecasts, and as we shall see could have serious consequences.

Through Gaia theory I offer a view of our and the Earth's possible future as climate change develops. It differs from that of most climate scientists. The differences come from procedure, not from a different factual basis. Most of the climate-change models, for example, do not yet include the physiological response of the ecosystems of the land or oceans. In no way is this the consequence of a battle between theories; it is that climate models stretch our mental and computing capacities so much that it takes a long time before new procedures can be included reliably – it is somewhat like changing the transport system of a city from buses to trams. In an ideal world all-inclusive climate models might lessen or even remove the disagreement but we cannot afford to wait for perfected models: we have to act now, so I offer predictions based on simple models from Gaia theory and evidence from the Earth now and in the past.

Professional climatology is based mainly on geophysics and geo-chemistry and often assumes that the Earth is inert and incapable of a physiological response to climate change. What makes the ideas in this book different is that they are based on a consistent theory of the Earth, Gaia, which has proven itself by successful predictions, and is beginning to be accepted as the conventional wisdom of Earth and life science. Do not suppose that conventional wisdom among scientists is similar to consensus among politicians or lawyers. Science is about the truth and should be wholly indifferent to fairness or political expediency.

When I criticize the IPCC consensus, I am most of all criticizing the lack of wisdom among managers and politicians who forced (I suspect unwilling) scientists to present the conclusions of different national and regional climate centres this way. Just before completing this book I read Steve Schneider's deeply moving recent work *The Patient from Hell*, about his long and painful but successful battle with cancer. Schneider is one of the world's leading climate scientists, and he recalls in the book his part in a session at the UN in Geneva during the development of the IPCC Working Group II report of 2001, describing how the good science presented at the session was manipulated until it satisfied all of the national representatives present. The book makes clear that the words used to express the consequences of global heating were blurred until they were acceptable to representatives from the oil-producing nations, who saw their national interests threatened by the scientific truth. If this is what the UN means by consensus, scientific truth cannot be expected to come from its deliberations and we are misled about the dangers of global heating. This may also be why national governments and international agencies are reluctant to fund observation and measurement but ready to fund models. Measurements by scientists are much harder to contest. It is said that truth is the first casualty of war and it seems that this is also true of climate change. If I am more right than the consensus, it alters profoundly the best course of individual and political action. Simply cutting back fossil-fuel burning, energy use and the destruction of natural forests will not be a sufficient answer to global heating, not least because it seems that climate change can happen faster than we can respond to it and it may be irreversible. Consider: the Kyoto Agreement was made more than ten years ago, and it seems that we have done little more to halt climate change since then other than almost empty gestures. Because of the rapidity of the Earth's change we will need to respond more like the inhabitants of a city threatened by a flood. When they see the unstoppable rise of water, their only option is to escape to high ground; it is too late for them to do anything else, as it is for us to try to save our familiar world.

The concept of a living Earth is not easy to grasp even as a metaphor. I will try to explain it later in this book but for now ignore dissimilarities such as that the Earth does not appear to reproduce. The evidence that the Earth behaves like a living system is now strong. It can either resist climate change or enhance it, and unless we take this into account we can neither understand nor forecast the Earth's behaviour. Keep in mind that it is hubris to think that we know how to save the Earth: our planet looks after itself. All that we can do is try to save ourselves.

Those of us who still walk in what was once the countryside sense that something is wrong or missing when we see a modern agri-business farm, with its fields full of monoculture crops, and we feel the same about dark gloomy forestry plantations of conifers sown in regimented ranks close together to maximize the quantity and quality of the timber and the foresters' gain. A few of us find it terribly wrong when some jewel of coastline or rural scenery is debased by plantations of giant industrial-scale wind turbines. Yet if we go to untouched forest, desert or indeed any place on Earth where things still grow in dynamic coexistence, we find it beautiful but scary, a place that alerts our sense of danger. The extrovert explorer in his bush hat will say, 'Nonsense, I have spent much of my life in the wilderness and never felt threatened.' He forgets that he also wears snake boots, and his kit carries water-sterilizing tablets and anti-malarial pills. Make no mistake, our instinctive fear of the wild is sound: wholly natural places are as inimical to innocent city folk as is the landscape of an alien planet infested by monsters. Forms of life from micro-organisms to nematodes to invertebrates to snakes, tigers and, of course, other humans: all these living things are potentially dangerous to us should we settle near them. No wonder early man set aside his fields from nature and gradually became a farmer, seeing all life other than livestock, crop plants, hired help and kinfolk as malign. Later we built cities – fortresses – to keep us safe from wildlife, and to overpower the countryside, making it serve our needs for food, fuel, minerals and building materials. There is nothing unnatural about this evolution. Termites and other social animals have done it in their way too. Where we differ from all that came before is that we escaped the causes of early death, predation, famine and disease, the things that once frightened us. Now we have multi-plied and expanded our cities, and so filled them that we overfill the Earth and make Malthus's nightmare real, despite our greatly increased capacity to sustain ourselves, which he had not foreseen. The natural world outside our farms and cities is not there as decoration but serves to regulate the chemistry and climate of the Earth, and the ecosystems are the organs of Gaia that enable her to maintain our habitable planet.

You think I exaggerate – but when did you last sit down on a warm grassy bank in the sun and smell wild thyme, or see the oxlip and a nodding violet? I bet it was a long time ago, if ever. Shakespeare could do it when he lived in London, because such a grassy bank was only a walk away from his home, and when I was a boy living in South London eighty years ago the tram would take me to such a bank in thirty minutes; now it is almost an impossibility. The city and its agricultural hinterland is nearly everywhere, and it is so large.

Should this seem a somewhat parochial English perception of the changing Earth, it is a matter of geography, not of tribal prejudice. As the climate crisis worsens the entire world will be affected, but in different ways. Sir John Houghton reminded us in his book *Global Warming*, published in 2004, that the greatest climate changes would be seen in the polar regions. First the floating ice will melt and then the ice caps of Greenland and Antarctica will erode; the consequences of these Arctic and Antarctic climate changes will be added heat and rising sea levels for the whole Earth, and then we will all feel the change. Except for those tropical places where mountains are close to a warm ocean and bring rain, greater heat means drought and a fatal loss of food production. Hot weather brings more rain but it runs off in flash floods or evaporates so quickly that it is far less use to growing plants than is the gentle drizzle that falls on a cool land like Ireland. For the continental areas where most of us live in the northern and southern hemispheres, the droughts of summer will intensify. In the USA they will revive memories of the Dust Bowl of the 1930s. Australia has already suffered eleven years of continuous drought; Europeans will recall the awful summer of 2003; and in China, Africa and Southern Asia famine is a familiar enemy. Like the foot of an elephant on an anthill, global heating will crush life from the continental plains.

What will it be like in a few years' time? We saw that in 2007 the Earth passed a significant milestone when the area of floating Arctic ice that melted in the summer was about three million square kilometres greater than usual, an area thirty times larger than England.

Despite the heat absorbed the global temperature did not rise, in fact it fell slightly, perhaps because to melt ice it takes eighty-one times as much heat as to raise the same quantity of water one degree: this property of ice is called its 'latent heat'. You can see this yourself by making a near full cup of tea with boiling water. It will be too hot even to sip. Adding cold water to cool it quickly rarely works, but add a single ice cube and it will be cool enough to drink in a few seconds. In a few more years all that floating ice may go and then the sun will be free to heat the dark Arctic Ocean. No longer will it have the Sisyphean task of trying to melt white reflecting ice that rejects 80 per cent of the sunlight it receives so that to melt it consumes most of the radiant energy that would otherwise warm the ocean. Keep in mind the fact that before the climate can return to its pre-industrial state all the melted ice has to be frozen again, and this means repaying the latent-heat debt of the ice. The American scientist Wally Broecker warns in his new book, *Fixing Climate*, written with Robert Kunzig, of potentially devastating global climate change consequent upon small changes in the Arctic climate

Some parts of the world may escape the worst. The northern regions of Canada, Scandinavia and Siberia, where not inundated by the rising ocean, will remain habitable and so will oases on the continents, mostly in mountain regions where rain or snow still fall. But the more important exceptions to this planet-wide distress will be the island nations of Japan, Tasmania, New Zealand, the British Isles and numerous smaller islands. Even in the tropics, global heating may not disable island communities such as those on the Hawaiian Islands, Taiwan or the Philippines. The British Isles and New Zealand will be among the least affected by global heating. Their temperate oceanic position is likely to favour a climate able to sustain abundant agriculture. They will be among the lifeboats for humanity. For the nations that occupy the continents, all may depend on their population density. The USA and the Russian states are singularly fortunate in having densities 8 and 30 times less than the UK respectively and both contain vast areas of previously frozen territory in their northern regions. The Indian subcontinent, China and South-East Asia, however, are fully populated and nations like Bangladesh are already threatened by rising sea level.

The human world of the lifeboat islands and continental oases will be constrained by limited food, energy and living space. The ethics of a lifeboat world where the imperative is survival are wholly different from those of the cosy self-indulgence of the latter part of the twentieth century. I cannot help wondering how we will manage – how we will decide who among the thirsty will be allowed aboard. We in the UK have little land left to farm and feed ourselves, but we and the refugees may in any case not be able to do so because the majority of us are urban, caring little for the world outside the city and not understanding that all our lives depend upon it. The high-sounding and well-meaning visions of the European Union of 'saving the planet' and developing sustainably by using only 'natural' energy might have worked in 1800 when there were only a billion of us but now they are a wholly impractical luxury we can ill afford. Indeed, in its way the green ideology that now seems to inspire Northern Europe and the USA may be in the end as damaging to the real environment as were the previous humanist ideologies. If the UK government persists in forcing through impractical and expensive renewable energy schemes we will soon discover that nearly all of what remains of our countryside becomes the site for fields planted with biofuel crops, biogas generators and industrial-sized wind farms – all this when what land we have is wholly needed to grow food. Don't feel guilty about opting out of this nonsense: closer examination reveals it as an elaborate scam in the interests of a few nations whose economies are enriched in the short term by the sale of wind turbines, biofuel plants and other green-sounding energy equipment. Don't for a moment believe the sales talk that these will save the planet. The salesmen's pitch refers to the world they know, the urban world. The real Earth does not need saving. It can, will and always has saved itself and it is now starting to do so by changing to a state much less favourable for us and other animals. What people mean by the plea is 'save the planet as we know it', and that is now impossible.

I think it unlikely that serious harm can come from the small-scale use of biofuels made from agricultural waste, recycled cooking oil or a modest harvest of ocean algae. But planting crops of sugar cane, beetroot, maize, oilseed rape and other plants solely for fuel production is almost certainly one of the most harmful acts of all. The trouble with mankind is, as William James said, 'Man can never have enough without having too much.' Once biofuel is used to keep our cars and trucks moving we will try to grow it globally, with appalling consequences. To give some idea of the scale already involved, consider the energy legislation enacted in 2007 in the USA, which envisages some $170 billion for biofuel refineries and their infrastructure. Brent Erikson, of the Biotechnology Industry Organization said, 'We are at the point where we were in the 1850s when kerosene was first distilled,' and he went on to say that the new law requires the production of 36 billion gallons of ethanol fuel from corn kernels by 2022. It is clear from Mr Erikson's statements, and from what is now happening in Brazil and from the European intentions, that biofuels are not some harmless cottage industry: they are big business as usual. How long before we become dependent upon biofuel to run our trucks and cars?

Does America understand the threat of global heating? Few would doubt that the USA is at present the top nation for science and invention, and what better proof than the computer that sits on all our desks and fulfils at the very least the work previously done by an audiotypist. The USA has played a major role in its evolution. If this alone were not enough, we have the landings on the Moon, the exploration of Mars and the fleets of amazingly intricate satellites from the Hubble telescope to those that tell you exactly where you are anywhere in the world. All this and much more is a tribute to American know-how and their can-do attitude. Even Gaia theory was discovered in the fertile environment of the Jet Propulsion Laboratory in California, and the one biologist who understood it and developed it further was that eminent American scientist Lynn Margulis. Of course, advances in science and technology emerged in Europe in the Middle Ages and moved its centre of excellence among the nations. In computer technology and theory Babbage, Ada Lovelace and that most tragic of men Alan Turing all did the groundwork here in the UK. Turing was the one who with his group built the first serious computing device and used it to deconvolute the otherwise unbreakable code of our wartime enemies. But that was then. Now America is at the centre of science.

I make this paean of praise to the United States of America because I am puzzled that, despite its scientific excellence, this of all nations was among the slowest to perceive the threat of global heating. I doubt that this unexpected ignorance is connected with the fact that the per capita American use of fossil fuel, one source of climate damage, is greater than anywhere else. I see it as more the consequence of most American scientists, in their straightforward successful and reductionist way, seeing the Earth as something that they could improve or manage; they seemed to see it as no more than a ball of rock moistened by the oceans and sitting within a tenuous sphere of air. They even seemed to see Mars as a planet to be developed when the Earth is no longer habitable. They do not yet see the Earth as a live planet that regulates itself.

They fail to see that because the Earth was colonized by life at least three and a half billion years ago, its temperature and surface composition have been set by the preferences of whatever organisms made up the biosphere. This was true in the cold of the ice ages, it is true now, and will be true in the heat of the hot age soon due. Of course the physics and chemistry of the air are important in the understanding of climate, but the manager of climates is and has always been Gaia, the Earth system of which the biosphere is a part. The disastrous mistake of twentieth-century science was to assume that all we need to know about the climate can come from modelling the physics and chemistry of the air in ever more powerful computers, and then assuming that the biosphere merely responds passively to change instead of realizing it was in the driving seat. Because we acknowledged the leadership of America in science, most of the world took its mistaken view as true. Almost too late, leading scientists worldwide are realizing that real observations and measurements falsify the twentieth-century view of the Earth as a passive resource. It may be good enough for weather forecasts but not to foresee the climate decades from now.

The quality of individual professional scientists in America is unmatched and it is they who are accurately observing the global environment: the names of Ralph Keeling and Susan Solomon come immediately to my mind but there are many others in the same rank in NASA, the National Oceanic and Atmospheric Administration

(NOAA) and the university science departments. America also re-deems itself though the powerful messages from Al Gore, Jim Hansen and Steve Schneider. Their words make us all aware how serious global heating is, but with the exception of E. O. Wilson, Stephen Schneider, Robert Charlson and a few other Earth scientists, the majority still shy away from the difficult concept of a live Earth. Our proper responses and actions to prevent, or more likely escape, the worst, still require that science embraces this concept and abandons the sterile ideas of mainstream Earth and life science. A change of view is emerging in the USA and may re-establish their leadership in this vital part of science.

Perhaps scientists should be conscripted to serve as was done in the Second World War, and by this I do not mean for something resembling the Manhattan Project alone. In the UK there was a tectonic shift in attitudes among scientists during the War. I well recall being interviewed for my first job as a new graduate in June 1941 at the National Institute for Medical Research, then in Hamp-stead. The interviewer was the director of the Institute, Sir Henry Dale; he was also President of the Royal Society and a recipient of the Nobel Prize. He was a formidable intellect and a kindly man with a very direct manner. Among his first words to me were, 'Put aside all thoughts of doing science here – science is suspended for the duration of the War; all we have to offer are ad hoc problems that need solving today or better yesterday.' He then added, 'After the War we will get back to real science and it will have been worth the wait.' Of course Sir Henry was wrong. The War provided fertile ground for real science when the slow pedestrian peacetime research was put aside. I found wartime science enthralling and stimulating and when peace came was appalled by the return to the pursuit of personal aggrandizement and the loss of the sense of wonder that so mars modern science. Remember, penicillin was first developed during the War and the whole concept of antibiotics born. Remem-ber also when you use your microwave that the magnetron at its heart was invented by Boot and Randal in the 1940s to improve wartime radar. Radar research led directly to radio astronomy and a new enlightenment about the universe. In Germany, the pressures for wartime invention led von Braun to develop the rockets that were the basis of the space science that enables us now to take Earth-orbiting satellites for granted and to see planetary exploration by robot vehicles as an affordable luxury.

Politicians of the developed world acknowledge climate change but their policies still appear to be those of the twentieth century, based on the advice of green lobbyists and those of the business community who see massive short-term gain from subsidized energy schemes. They rarely seem to act on the recommendations of their science advisers. At Bali, political leaders agreed to cut all carbon emissions by 60 per cent by 2050. What on Earth made them think that they could make policy for a world over forty years ahead? Policies based on unjustifiable extrapolation and environmental dogmas are unlikely to avert climate change and we should not even try to implement them. Instead our leaders should immediately concentrate their minds on sustaining their own nations as a viable habitat; they could be inspired to do this not just out of selfish national interest but as captains of the lifeboats that their nations might become. In early 2008 the UK government at last announced a programme for building new nuclear power stations. I surely hope that this is not another of the false promises that characterized so many of the utterances of the Blair administration. Nuclear energy is by far the most effective way to reduce the emission of carbon dioxide, but that is not the most important reason for us to emulate France and make electricity from uranium. What is important is that cities demand a constant and economic supply of electricity which until recently has come from coal and gas but these are now declining and leave no alternative to nuclear energy. Huge flows of electricity will be demanded by the mega-cities that are starting to emerge and this can only be met in the near term by a vigorous and rapid expansion of nuclear energy. This need is intensified by the fact that we have little land on which to grow food and intensive agriculture needs abundant energy. As oil runs out we will need to synthesize fuel for the mobile machinery of construction, transport and agriculture. This is not difficult to do from coal or nuclear energy but we need to start preparing for it now. We may even have to consider the direct synthesis of food from carbon dioxide, nitrogen, water and tissue culture.

There will be a flood of anti-nuclear energy disinformation from energy companies whose profitability is threatened and even from nations who see their power and influence lessened. Do not believe such tales as that the construction of a new nuclear power source takes ten to fifteen years. Construction takes the French less than five years and there is no reason why we should take longer, if we avoid the excessive time spent in planning agencies, courtrooms and at public hearings. I hope that the green movement and their attendant lawyers do not continue their mistaken opposition to nuclear energy. Most of it is irrational and based on an unsustainable concatenation of mistakes and misinformation that are amplified by the media. It would be good if journalists and editors tempered their desire to tell a scary story with the reality that without an ample supply of nuclear energy life on our islands may in one or two decades decline to a state of poverty. By putting humanity first, and neglecting Gaia, too many greens have sown the seeds of their own destruction and, if they persist, ours as well; they could mitigate their error by dropping their delaying tactics against nuclear energy. More importantly, they would then be helping to power the lifeboat not, as now, sabotaging its engine.

It is absurd to think that we in the UK can alter the Earth's response in our favour by using wind or solar voltaic energy. A wind farm of twenty 1 MW turbines requires over 10,000 tonnes of concrete. It would require 200 of these wind farms covering an area the size of Dartmoor to equal the constant power output of a single coal-fired or nuclear power station. Even more absurd, a full-sized nuclear or coal-fired power station would have to be built for each of these monster wind farms to back up the turbines for the 75 per cent of time when the wind was either too high or too low. As if this were not enough to damn wind energy, the construction of a 1 GW wind farm would use a quantity of concrete, 2 million tons, sufficient to build a town for 100,000 people living in 30,000 homes; making and using that quantity of concrete would release about 1 million tons of carbon dioxide into the air. For us to survive as a civilized nation our cities need that safe, secure and constant supply of electricity that only coal, gas or nuclear can provide, and only with nuclear can we be assured of a constant supply of fuel. We have already seen how vulnerable gas supplies are to the continued integrity of pipelines perhaps a thousand miles long, and to the aggressive politics of autocrats. Coal is expensive in the UK and imports are insecure. Wind farms are hopelessly inadequate to the UK as a source of energy and as I have indicated can do little to halt global heating even when used on a global scale; moreover, experience in Western Europe shows them to be costly and inefficient sources of electricity. You will soon discover this when your electricity bills and taxes rise to pay for renewable energy we do not need. Your money will provide easy profits to be taken from the subsidy trough. These bills are imposed upon us so that politicians can appear green and good and some European nations grow richer. It does nothing for the Earth and will only add more stress for our island nation, and perhaps lead to its final breakdown.

The most frequent response from my green friends to the grim message in my last book was: 'You can't say things like that. It gives us nothing to hope for.' This was a good criticism, which helped to clear my mind and let me understand why messengers are said to have a short lifespan. I realized that I had said much about the imminent catastrophe but too little about how we could try to ensure our continued presence on the Earth, giving our descendants a chance in the hot world that soon may come. We are the intelligent elite among animal life on Earth and, whatever our mistakes, Gaia needs us.

This may seem an odd statement after all that I have said about the way twentieth-century humans became almost a planetary disease organism. But it has taken Gaia 3.5 billion years to evolve an animal that can think and communicate its thoughts. If we become extinct she has little chance of evolving another. I will enlarge this thought later in the book.

When I am warned that my pessimism discourages those who would improve their carbon footprint or do good works such as planting trees, I'm afraid I see such efforts as at best romantic nonsense, or at worst hypocrisy. Agencies now exist which allow air travellers to plant trees to offset the extra carbon dioxide their plane adds to the overburdened air. How like the indulgences once sold by the Catholic Church to wealthy sinners to offset the time they might otherwise spend in purgatory. Thirty years ago I foolishly planted 20,000 trees, hoping to restore to nature the farmland I had bought. I now realize it was a mistake: I should have left the land untouched and let an ecosystem, a natural forest, emerge filled with biodiverse and abundant life, in Gaia's own time. Unlike a mere plantation, such a forest could evolve, or die if it had to, as the climate changed. Planting a tree does not make an ecosystem any more than putting a liver in a jar fed with blood and nutrients makes a man.

I hope that Tim Flannery's fine book *The Weather Makers*, and my last book, *The Revenge of Gaia,* achieved some of their purpose. Both were intended as wake-up calls like that once-heard cry of pub landlords, 'Last orders. Time, gentlemen, please!' – a warning that soon the doors would close and we would be cast into the weather outside. I hope that a sufficient number of us are now aware that the lush and comfortable world that once we knew is departing for ever. But I fear that we still dream on and rather than waking we weave the sound of the alarm clock into our dreams.

Perhaps we are not aware of how fast the world is changing because we are so adaptable. If the average temperature in the UK in January is 7°C it feels cold most of the time, and we wrap up on frosty mornings or when a bleak north-west wind blows. We mutter, where is the global warming now? In the summer it averages 20°C in July and we enjoy a week with maximum temperatures of 30°C, but grumble if it falls to 15°C for as long. Yet only twenty years ago, these winter and summer temperatures would have registered as unseasonably warm. The rainfall of the United Kingdom's eastern counties has always been low, in the region of 20 inches per year, but the countryside was always lush and green, because it stayed cool during the summer. By contrast, Arizona, which has a similar rainfall, is almost entirely scrub and desert simply because it is so much hotter, and the rain that falls dries up or runs off into channels before the plants can benefit. Our most south-easterly county, Kent, is already growing short of water, and Southern Europe is now almost a desert. Adapting, as individual animals, is not so hard to do: when a tribe moves from temperate to tropical regions it only takes a few generations before individuals darken as selection eliminates the fair-skinned. So it is with all of us: our world has changed for ever and we will have to adapt, and to more than climate change. Even in my lifetime, the world has shrunk from one that was vast enough to make exploration an adventure and included many distant places where no one had ever trod. Now it has become an almost endless city embedded in an intensive but tame and predictable agriculture. Soon it may revert to a great wilderness again. To survive in this new world we need a Gaian philosophy, and to prepare ourselves to fight a barbarian warlord out to seize us and our territory.

Apart from the occasional disastrous flood, excessive heat wave, or wholly unexpected frost, the climate in the UK will change slowly and imperceptibly at first. People in cities like London will forget that, even in days of wealth not so many years ago, air-conditioning was almost never needed in the summer, whereas my colleague Gari Owen reminds me that London in 2006 used more energy for cooling than for heating. In the short term, nothing much is likely to happen with the climate here that would stir a rebellion. What might do so are the disastrous consequences of sea level rise leading to the destruction of a major city or the failure of food or electricity supplies. These dangers will be aggravated by the ever-growing flux of climate refugees, to which will be added returning expatriates who left the crowded United Kingdom for what they thought would be a pleasant life in Europe. Our gravest dangers are not from climate change itself, but indirectly from starvation, competition for space and resources, and tribal war.

In a small way the plight of the British in 1940 resembles the state of the civilized world now. At that time we had had nearly a decade of the well-intentioned, but quite wrong belief that peace was all that mattered. The followers of the peace lobbies of the 1930s resembled the green movements now; their intentions were more than good, but wholly inappropriate for the war that was about to start. The fundamental flaw of the green lobbies now is revealed in the name Greenpeace; by conflating the humanism of peace movements with environmentalism they unconsciously anthropomorphize Gaia. It is time to wake up and realize that Gaia is no cosy mother that nurtures humans and can be propitiated by gestures such as carbon trading or sustainable development. Gaia, even though we are a part of her, will always dictate the terms of peace. Back in May 1940 we woke to find facing us across the Channel a wholly hostile continental force about to invade. We were alone without an effective ally but fortunate to have a new leader, Winston Churchill, whose moving words stirred the whole nation from its lethargy: 'I have nothing to offer but blood, toil, tears and sweat.' We need another Churchill now to lead us from the clinging, flabby, consensual thinking of the late twentieth century and to bind the nation into a single-minded effort to wage a difficult war. We need a leader who will stir us all but especially to stir those young green activists who so bravely protested against all forms of desecration of the countryside. Where are the 'Earth First' battalions, and where have Swampy and his friends gone?

What has most moved me during the writing of this book is the thought that we humans are vitally important as a part of Gaia, not through what we are now but through our potential as a species to be the progenitors of a much better animal. Like it or not we are now its heart and mind, but to continue to improve in this role we have to ensure our survival as a civilized species and not revert into a cluster of warring tribes that was a stage in our evolutionary history. I am stirred by the thought that the Earth system, Gaia, has existed for more than a quarter of the age of the universe and it has taken this long for a species to evolve that can think, communicate and store its thoughts and experiences. As part of Gaia our presence begins to make the planet sentient. We should be proud that we could be part of this huge step, one that may help Gaia survive as the sun continues its slow but ineluctable increase of heat output, making the solar system an increasingly hostile future environment. We have to do all that we can, and Chapter 5 is about the ideas now circulating among scientists and engineers that might reverse climate change. So far they are untried, uncertain, and conceivably dangerous, some-what like medicine and surgery in the nineteenth century. If we can keep civilization alive through this century perhaps there is a chance that our descendants will one day serve Gaia and assist her in the fine-tuned self-regulation of the climate and composition of our planet.

We have enjoyed 12,000 years of climate peace since the last shift from the glacial age to an interglacial one. Before long, we may face planet-wide devastation worse even than unrestricted nuclear war between superpowers. The climate war could kill nearly all of us and leave the few survivors living a Stone Age existence. But in several places in the world, including the UK, we have a chance of surviving and even of living well. For that to be possible we have to make our lifeboats seaworthy now. Even if some natural event such as a series of large volcanic eruptions or a decrease of solar radiation reprieves us, it still will have been better to spend our money and our efforts making our countries self-sufficient in food and energy and, if we are to become wholly urban, then in making cities that we are proud to live in.

2

The Climate Forecast

In the last chapter I made some strong statements about the future climate and its consequences for us all. You may well ask with what authority I make them. Why should you read – much less believe – a lone scientist, when the consortium of most of the world's climate professionals, the IPCC, appears to express a much milder consensus on climate change. My qualifications are listed in my autobiography, *Homage to Gaia*, but what makes my forecast of future climate different is not simply a matter of disagreement among scientists, although that is normal and healthy enough; why I speak out so strongly and talk of catastrophe is because I am a scientist influenced by evidence coming from the Earth, and viewed through Gaia theory. I work independently and I am not accountable to some human agency – a religion, political party, commercial or government agency. Independence allows me to consider the health of the Earth without the constraint that the welfare of humankind comes first. This way I see the health of the Earth as primary for we are utterly dependent upon a healthy planet for survival. What inspired me to write this book was hearing in the autumn of 2007 that the IPCC had reached a consensus on future climate. I know and respect the scientists of the IPCC and several of them are personal friends but I was shocked to hear that they had reached a consensus on a matter of science. I know that such a word has no place in the lexicon of science; it is a good and useful word but it belongs to the world of politics and the courtroom, where reaching a consensus is a way of solving human differences. Scientists are concerned with probabilities, never with certainties or consensual agreement. (Lewis Wolpert's book *The Unnatural Nature of Science* is a fine introduction to these different ways of thinking.) The IPCC is potentially the most effective link we have between climate science and human affairs and policy: it was bad enough to see an honest probabilistic set of model predictions presented as a consensus, but when in addition I saw how greatly climate observations in the real world differed from the model forecasts of the IPCC made a few years ago, I knew that I had to speak out.

Moreover, it seemed that there was little understanding of the great dangers we face. The recipients of the climate forecasts, the news media, government departments, the financial market – normally as skittish as blushing teenagers – and the insurance companies all seem relatively unperturbed about climate change and continued with business as usual until their world, the global economy, almost collapsed. Indeed the only noticeable change to normal life is the ever-growing urge to appear green, made more difficult by the straitened circumstances brought on by an incipient recession.

I wish that I had more confidence in our capacity to forecast the climate of 2050. I remember too well the forecasts of the present day made in the 1960s. None of them even glimpsed the climate changes that have already happened; indeed most thought an ice age more likely than global heating. The best guesses at 21st-century life shared the vision of that great prophet Herman Kahn, who foresaw a benign hi-tech world with everyone living at the standard of Scarsdale, the suburb he lived in near New York, and if you look at the burgeoning middle class of India and China now he was not far wrong. Kahn was good at predicting the way the human world would go but wholly ignorant about the Earth and the consequences of the rapid growth of population, agriculture and energy-intensive industry. As confident as Kahn, our politicians now talk with conviction about a world in 2050 fit for 8 billion people living on a 2°C hotter Earth with the temperature stabilized and emissions regulated. I wonder if an Intergovernmental Panel on Economic Change will be as sanguine about 2050. We deplore the clever manipulators who 'trash and cash' by short-selling a bank, but praise governments who provide subsidies for the snake-oil remedies for climate ills and easy money for the firms that sell them. We still seem to think that by mid century we will enjoy a well-run and comfortable world under human management and stewardship. In the 1960s we were wholly unaware that we inhabit a live planet whose needs are in conflict with our own. It is too easy to make guesses about the future when we all imagine that life will be much the same as now but with a few interesting or unpleasant details bolted on. That is why Kahn was so successful. His clear message was: carry on business as usual and all will be well – just what we wanted to hear. I see no significant change between his recommendations and those of our politicians and their advisers now. To be sure they hedge their bets with green-sounding invocations and aim for sustainable development, but can this do more than the prayers they offer in Parliament?

I am not a contrarian; instead I greatly respect the climate scientists of the IPCC and would prefer to accept as true their conclusions about future climates. I do not enjoy argument for its own sake but I cannot ignore the large differences that exist between their predictions and what is observed.

In human affairs we know that 'he who hesitates is lost'; social scientists talk of 'cognitive dissonance', which the composer of the phrase, Leon Festinger, defined as the feeling of discomfort we feel when trying to hold two contradictory ideas simultaneously and the urge to reduce the dissonance by modifying or rejecting one of the ideas. It operates when we choose between two almost equal objects and, having chosen, invest our choice with superlative advantage over the alternative so that we can happily reject it. The decision process must be part of our genetic inheritance; we need that certainty in human transactions. We have to choose and then have faith in our choice; this applies to the jobs we take, how we vote, the purchases we make, and the marriages to which we commit ourselves. It applies also to a judge or jury, but it is worse than useless in science. However, scientists are human and we never entirely escape the pull of cognitive dissonance.

The range of forecasts by the different models of the IPCC is so large that it is difficult to believe that they are reliable enough to be used by governments to plan policy for ameliorating climate change. It is a brave try at an exceedingly difficult scientific task and probably we are expecting too much from them: it would be wrong to expect the view of the panel to be truly authoritative. The main reason for doubt is the fact that the forecasts do not agree with high-quality evidence from the Earth obtained by scientists whose job it is to measure and observe. This evidence reveals the failure of the IPCC to forecast correctly the course of climate change up to 2007, and I will present it in detail shortly. Moreover, the long-term climate history of the Earth reveals the existence of several stable but quite different climate states, and their existence is not predicted by present-day climate models. I trust the observations of the scientists who make measurements of the climate, and equally the all-seeing impersonal observations of the satellites that tirelessly view the Earth from space and the automated ocean observers that continuously report the state of the waters. But I have much less confidence in the models that forecast future climates. We should not expect climate models to be reliable – they have only recently evolved from the short-term needs of weather forecasting and are limited by a climate theory based almost wholly on atmospheric physics, and even this is far from complete. The science is good within its limitations but a full understanding of the climate involves much more than atmospheric physics alone. Discerning scientists at several of the major climate centres are making serious efforts to build climate models that are more comprehensive, but surely it is unwise of governments to base policies looking more than forty years into the future on forecasts made several years ago by models acknowledged to be incomplete.

What is the evidence for thinking that the IPCC may be under-estimating the severity of climate change? In May 2007 a single-page paper was published in *Science* by a group of authors, all prominent climate scientists (Rahmstorf et al.). Figure 1 is taken from this paper to illustrate their findings.

In the lower panel the broad grey zone indicates the IPCC pre-dictions of sea level rise up until 2007 and the upper solid line and set of connected points represents the average and individual measurements of sea level from 1970 until 2007. The measured sea level has risen 1.6 times as fast as was predicted. Similar but lesser discrepancies exist with the predictions of temperature and these are shown in the upper panel. The shaded zone is again the IPCC's range of predictions and the wandering line joining the crosses (x) the observed global mean temperature. The discrepancy is not as

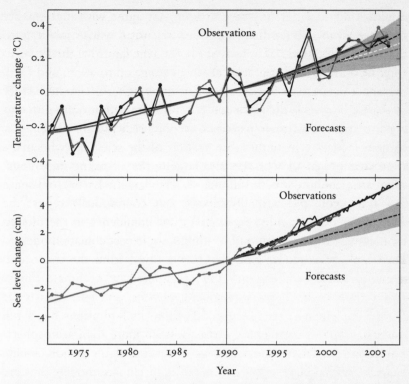

Figure 1. *Upper panel: comparison of observations of global mean temperature (joined points) with model forecasts (grey zone and dotted lines). Lower panel: comparison of observed sea level (joined points) and model forecasts (grey zone and dotted lines). Both panels cover the years 1970 to 2007.*

great as that for sea level but still serious considering we are comparing the predictions with what has actually happened.

To me the most important quantity is not the global mean temperature, but how much extra heat the Earth has absorbed from the sun. The global mean temperature is like the current account balance, which inevitably varies from day to day; the total heat absorbed is an indication of the reserves.

Sea level rise is the best available measure of the heat absorbed by the Earth because it comes from only two main causes: the melting of glaciers on land and the expansion of the ocean as it warms – in other words, sea level is a thermometer which indicates the true global warming. Look at Figure 1 again and note how the sea level progressively rises but the mean global temperature fluctuates year by year. Schneider tells me that a similar but more regionally discerning measure of the total heat absorbed is the height of the atmosphere. Like the ocean, the air expands as it warms.

The next contrary evidence comes from observations of the area of the Arctic Ocean covered in summertime by floating ice. In 1980 and previous years the area covered at the end of September (when ice cover is least after a summer of melting) showed 10 million square kilometres of ice, an area about as large as the USA. In 2007 it had fallen to 4 million square kilometres. Figure 2 compares the range of IPCC predictions with the observed decline of floating ice. The discrepancy is huge and suggests that if melting continues at this rate the summer Arctic Ocean will be almost ice-free within fifteen years. The IPCC prediction suggests that this is unlikely before 2050.

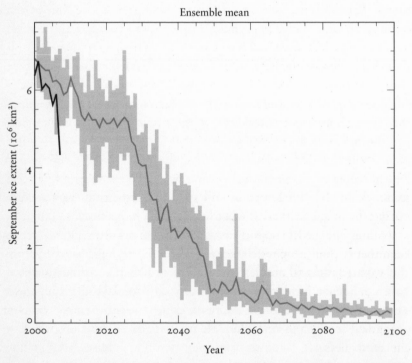

Figure 2. *The IPCC model predictions of the extent of ice covering the summer Arctic Ocean (grey zone with a solid line representing the average in its centre) and the observed ice cover (solid line to the left of the figure).*

The melting of floating ice does not significantly raise sea level, as Archimedes could have advised had he been asked, but it does make a difference to the quantity of heat received by the Earth from the sun. White, snow-covered ice reflects 80 per cent of the sunlight back to space but dark sea water reflects only 20 per cent of the sunlight that strikes it. The extra heating of the Arctic basin if all the floating ice melted would be 80 watts per m², which averaged over the whole Earth is an increase of one watt per m². This is a truly serious increase of the Earth's heat load. To put it in perspective, the extra heat which will be absorbed when the floating ice has gone is nearly 70 per cent of the heating caused by all of the carbon dioxide pollution now present.

The third piece of evidence comes from an article by Jeffrey Polovina in *Geophysical Research Letters*, published in 2008. He and his co-workers reported satellite observations of the ocean areas of the Earth from space that showed a progressive decline in the population of ocean algae. The authors comment that the barren area of ocean has increased by 15 per cent in the past nine years and that this is a consequence of global heating that has made the surface waters warmer and less well mixed with nutrient-rich waters below. Algal growth acts to cool the Earth by several mechanisms, including the removal of carbon dioxide from the air, so that warming reveals yet another positive feedback on global heating. In a 1994 *Nature* paper, the American scientist Lee Kump and I made a geophysiological model of this phenomenon and warned of its inherent positive feedback on global heating. So far as I am aware this phenomenon is not yet included in their models by climate professionals. But it is now an established fact of observation and another prediction from Gaia theory that has passed its test.

If we are failing to predict what has already happened how can we have confidence in predictions for forty or ninety years from now? Yet political action and governmental initiatives to combat climate change all seem to assume that the IPCC is at least making reliable educated guesses.

In addition to the discrepancies between modelling and observations, Gaia theory predicts a different course of climate change consequent on carbon dioxide pollution. This theory is acknowledged, but not yet used in practice by climate scientists, usually because they are not yet ready for it; in some ways they are like students of mathematics who realize the value of calculus but are not yet trained to use it. As a consequence climate scientists, even when they acknowledge the Earth's vitality, still act as if it were a dead planet like Mars or Venus because such planets are so much easier to model.

Professional scientists are usually specialists trained in a specific discipline or groups of disciplines. Nearly all of climate science is in the territory of atmospheric physics. These physicists sail huge climate models residing in powerful computers, as large and unwieldy in many ways as the iron-clad battleships of past centuries. Fortunately the ships' captains are brave meteorologists and have proven themselves at one of science's sharp ends, weather forecasting. Few other scientists have their mistakes subject to so discerning and public a post-mortem as weather forecasters. You all know how impenetrable is the set of programs running your desk computer or laptop. Just imagine what a general circulation climate model must be like if it needs a computer one thousand times more powerful than the one on your desk.

It is as easy to be lost in the vast intricacies of computer models as it is in a battleship. I once had the misfortune to be lost in the bowels of a large warship and remember with dismay the endless passageways, watertight doors and vertical ladders that linked the compartments of the ship's interior. These ships have become three-dimensional mazes so intricate that it is rumoured that dubious denizens have made within them no-go areas for themselves. The IPCC model fleet led by Admiral Pachauri sails forth on a wholly alien sea. As is often the way in wars, their battleships (being models of the climate war) are already outdated but not yet complete; the constructors are still aboard and some of them mutter, 'They really should have built something quite different,' but there is little that they or the Admiral can do. The atmosphere, whose physics they model, is not some simple gift of the Earth's geological past; it is, apart from about 1 per cent of the so-called rare or noble gases, entirely the product of living organisms at the surface. Much worse, these organisms, and that includes humans, are able to change their inputs and outputs of gases without letting the Admiral know.

Today's allies, the micro-organisms of the soil and ocean who help to cool the climate, can become tomorrow's enemies, and add carbon dioxide instead of removing it. Moreover, the air normally, that is before we humans started changing it, was kept dynamically at a constant composition and one that sustained a habitable climate.

The background planetary science that should be the basis for professional climatology has been in a state of flux and conflict for at least two hundred years. Many natural philosophers from the nineteenth century and earlier realized there was a link between life and the material Earth but even Erasmus Darwin, T. H. Huxley and V. Vernadsky did not go much further than anecdotal speculation. Proper science, which is the building of testable hypotheses, did not happen in this field until the twentieth century when the recognition of links between life at the surface and in the ocean led the great but insufficiently recognized scientists G. E. Hutchinson, A. C. Redfield and Lars Sillen to research the biochemistry of the ocean and land surface. And, quite naturally, they called their science biogeochemistry and established it as a separate discipline, and one now prominent in European science. It is important to note that biogeochemistry, like biochemistry in medical science, is not a systems science. It is not a physiology of the Earth and few biogeochemists would be happy to think of the Earth as in any way alive. For those unfamiliar with the niceties of scientific naming, the last science in a compound name of that kind is usually the leader, thus biochemists and biogeochemists are chemists by training who are working with life and its products, and biophysics is done by physicists working on biological subjects.

Geophysiology, the discipline of Gaia theory, had its origins in the 1960s Gaia hypothesis. Geophysiology sees the organisms of the Earth evolving by Darwinian natural selection in an environment that is the product of their ancestors and not simply a consequence of the Earth's geological history. Thus the oxygen of the atmosphere is almost wholly the product of photosynthetic organisms and without it there would be no animals or invertebrates, nor would we burn fuels and so add carbon dioxide to the air. I find it amazing that it took so long for biologists even grudgingly to acknowledge that organisms adapted not to the static world conveniently but wrongly described by their geologist colleagues, but to a dynamic world built by the organisms themselves.

Because of this convenient and very human separation of Earth's climate problem into the provinces of separated specialities, hardly any scientist sees it as a whole topic involving all the Earth, including humans, living organisms, the ocean, atmosphere and surface rocks. Most of all the separation prevents them seeing the Earth as a dynamic interactive system or, as I would put it, somehow alive. Scientists, you see, would rather go on doing what they were trained to do as specialists – their business as usual – than go back and take on the near impossible task of learning at least two other major branches of science. It would not be so bad were there more general practitioners to interpret between them.

The eminent climate scientist James Hansen, head of NASA's Goddard Institute for Space Studies in New York, has called for a far greater reduction in carbon dioxide than recently proposed by the European Union (EU). Hansen stated that the 550 parts per million upper limit proposed by Europe is far too high and that it should be as low as 350 ppm if humanity wishes to keep a planet similar to the one on which civilization developed. His strong statements are based on recent observations and on the Earth's climate history and, although he does not say it explicitly, I think he realizes that models based on atmospheric physics alone are unable to predict the future climate.

A simple model (illustrated in Figure 3) based on Gaia theory confirms Hansen's view and suggests a very different climate progress to that of the IPCC.

The figure illustrates the predictions of temperature change on a simple model planet. The planet is assumed to orbit the sun at the same distance as the Earth and is inhabited by two main ecosystems, algae in the oceans and plants on the land. The change of temperature as carbon dioxide abundance increases is determined in the model by the equations linking plant and algal growth with temperature and equations linking their presence with the carbon dioxide and clouds in the atmosphere. Geophysics as well as biology is equally important in the model. In particular, ocean physics determines that warm surface water separates and floats on the cooler

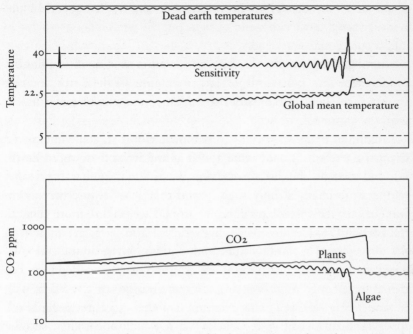

CO2 input going from pre-industrial to three times pre-industrial in one Kyr

Figure 3. *Climate change on the model planet described in the text as the carbon dioxide of its atmosphere is progressively increased. The lower panel illustrates the changes in ocean algae and land plant area and the carbon dioxide abundance. The upper panel shows the global temperature change in degrees Celsius and the sensitivity of the model (the rate at which the temperature increases as carbon dioxide is increased).*

waters below at temperatures greater than about 12°C and so denies the algae the nutrients needed for growth. On the land at temperatures above about 24°C rainwater evaporates rapidly enough to leave the land dry in between rainstorms. These two physical properties of bulk water set the upper temperature limit to plant and algal growth in their local environments. The model is a mix of highly non-linear differential equations but the strong feedbacks both negative and positive that link the biosphere to the atmospheric composition and climate constrain the evolution of the model and entirely avoid excursions into chaos; even so it is a dynamic not an equilibrium model. In this context something dynamic is alive whereas something at equilibrium is dead, like the difference between a living person and a stone statue. Both can stand upright but the person does so actively and would collapse if dead.

Although very simple, this kind of model is used by climatologists to diagnose the behaviour of larger climate models, but the one shown is unusual in the extent to which life on the planet is linked with the climate.

I made an experiment with this model world to see what would happen if carbon dioxide were added as we are now doing to Earth. I found that as the carbon dioxide was added, at first the global temperature changed only slightly and this was because the system was in negative feedback and resisted the perturbation, but as the carbon dioxide abundance approached 400 parts per million in the air, signs of instability appeared, shown by the amplification of small temperature fluctuations. It is important to recognize that dynamic self-regulating systems such as you, me or the Earth will, if sufficiently stressed, change from stabilizing negative feedback to destabilizing positive feedback. When this happens they become amplifiers of change. As amplifiers they do not distinguish between heating and cooling so that a small decrease of heating has a larger effect than might be anticipated and can cause a noticeable drop in temperature. Then suddenly, between 400 and 500 ppm of carbon dioxide, a small increase of heat or carbon dioxide causes a sudden 5°C rise of temperature. After that the model planet stabilizes again and resists further increases of carbon dioxide. The Earth's atmospheric greenhouse is now well above 400 ppm (carbon dioxide is near 390 ppm but methane, nitrous oxide and the CFCs lift the total effect to nearer that of 430 ppm carbon dioxide).

The experiment also included the sudden removal of all the added carbon dioxide soon after the hot state had been reached. Despite the carbon dioxide falling to 280 ppm the model planet stayed in its hot state. If this model resembles the real Earth then it suggests that stabilization is only possible on Gaia's terms at five degrees hotter than now, or at the previous stable climate about 200 years ago in pre-industrial times, or at the seven degrees cooler of a glaciation.

This model experiment also raises the question of the validity of the constant called sensitivity, used in nearly all large climate models, such as those of the IPCC. It is defined as the rise of temperature when the carbon dioxide in the air of the model is doubled. Mathematics allows sensitivity to be a constant if the equations of the model are linear; the recondite term 'non-linear' implies that properties such as temperature are not directly proportional to others, such as carbon dioxide, but are linked in ways that change with change. In the real world and in the simple model of Figure 3 the connections between climate and growth are usually quite non-linear. This non-linearity is the cause of the transition from negative to positive feedback as we approach the critical point of a temperature jump; at this point sensitivity, as the figure illustrates, is no longer constant but fluctuates. It is constant only at steady-state dynamic equilibrium or in the artificial equilibrium of a linear model.

Look closely at the upper panel of Figure 3 and note how the sensitivity drops to a minimum just before the temperature makes its jump to the stable hot state. The same effect is seen although less markedly in the temperature. If it truly represents the Earth's response to increasing carbon dioxide, it is scary because it implies that, before the final jump to a desert world, the climate will briefly become cooler again. This warns that a cold summer, or even a series of them, is not proof that global heating has ended.

These are some of the reasons that make me doubt the wisdom of applying the consensus of the IPCC to policies so far in the future. In addition to these reasons for doubt, all based on direct evidence from the Earth, I would like to focus tightly on another question important in climate modelling and prediction. What is the effect on climate of clouds and aerosols in the air?

Most of you reading this book have sat in the window seat of an aircraft and looked down at the land below. On a fine day when you look up from ground level you often see a clear and cloudless blue sky but soon after taking off if you look down you will see a white haze slightly obscuring the land below. This is the all-pervasive atmospheric aerosol that reflects sunlight back to space and makes global warming less severe than it might otherwise be. This haze comes mostly from pollution by cars, industry and farming but some comes from the algae in the ocean, and over the great oceans of the southern hemisphere the gaseous products of ocean life and dust blown from the deserts are the main source of the haze. Clouds in the air, when they are near the ground, reflect sunlight as do aerosols, but high-level clouds, for example the cirrus that heralds an approaching depression or the contrails of jet aircraft, add to global warming. Lastly, the haze and the clouds affect each other. Haze in humid air becomes cloud and the brightness of clouds is increased by haze particles; clouds can also hasten the removal of haze from the air.

In 2004 two IPCC contributors, Peter Cox and Meinrat Andreae, raised the question: what happens to global warming if this pollution haze suddenly disappears? Their paper in *Nature* warned that if the haze disappeared global heating would intensify and dangerous change could be the consequence.

In 2008, a group led by Peter Stott, from the Hadley Centre (part of the Metereological Office), examined this phenomenon in a careful and well-drawn paper in the journal *Tellus*: 'global dimming', they revealed, is complex even as a purely geophysical problem. According to their calculations the sudden removal of haze could lead to either a modest or a severe increase of heating. I now begin to see why my wise friend Robert Charlson is so loath to commit himself on pollution aerosols and climate change. Even so there was little doubt among any of these distinguished climate scientists that the present pollution haze reduces global heating, or that its sudden removal could have serious consequences.

I suspect that we worry less about global heating than about a global economic crash, and forget that we could make both events happen together if we implemented an immediate global 60 per cent reduction of emissions. This would cause a rapid fall in fossil fuel consumption and most of the particles that make the atmospheric aerosol would within weeks fall from the air. This would greatly simplify prediction and we could at last be fairly sure that global temperature would rise; the removal of the pollution aerosol would leave the gaseous greenhouse unobstructed and free at last to devastate what was left of the comfortable interglacial Earth. Yes, if we implemented in full the recommendations made at Bali within a year, far from stabilizing the climate, it could grow hotter not cooler. This is why I said in *The Revenge of Gaia*, 'We live in a fool's climate and are damned whatever we do.'

As if this were not enough, the American scientist Professor V. Ramanathan has recently drawn attention to the huge output of smoke and other aerosols from the fast-growing industries of Asia. The smoke cloud from China now stretches across the Pacific Ocean to North America and the sunsets in California have a rosy tinge redolent of a similar light-scattering by the stratospheric haze from the volcano Pinatubo that erupted in 1991. The same light-scattering is occurring over the Indian Ocean as India expands its industry.

These are new aerosol additions to the atmosphere: the smoke clouds from North America and Europe have moved similar distances for several decades across the Atlantic Ocean and into Asia. In addition to this, global smog from industry, smoke from burning forests in Africa, South America and from wildfires in the boreal forests of Canada and Siberia are all adding their ingredients to the witches' brew that the atmosphere has become.

Ramanathan has alerted us to the fact that these new pollution clouds are considerably darker than their predecessors from the US and Europe. They contain soot that absorbs sunlight, whereas the lighter aerosols mainly reflect sunlight. This makes the assessment of their effect on climate even more difficult.

The atmospheric physics of the connection between aerosols and climate is at the edge of comprehensibility and it is inevitably further confused by the feedbacks from other parts of the system. Clouds in the air are affected by life at the surface: airborne particles made by bacteria induce water droplets in clouds to freeze at temperatures as high as 2°C; otherwise super-cooled water droplets can cool to −40°C before they freeze. When they do freeze the heat released lifts the clouds and brings rain and thunder. In so very many different ways living things affect the climate as well as being affected by it: forests evapotranspire huge volumes of water vapour (evapotranspiration is an active physiological process by which ground water is pumped to the leaves); and ocean algae make gases that become the nuclei of cloud droplets. All that we have now are uncertain numbers attached to the numerous separated parts of the system and guidance from Gaia theory: we are like a nineteenth-century physician trying to give a sensible prognosis to a patient with diabetes. We can only make vague generalizations about the future, and were it not for the large and unquestioned greenhouse-heating effect of carbon dioxide, methane and other gases we would indeed be in the dark. A splendid account of our understanding so far of this complex science is in Robert Charlson's chapter of the book *Earth System Science*, published in 2001. To me, the message from aerosol and cloud research is that the global heating already experienced would be more severe without their presence, therefore we need to commit climate funding for monitoring and research.

Climatologists sometimes seem to think that the temperature of the leaves of a forest canopy can be calculated simply from knowledge of the albedo, that is the proportion of sunlight reflected back, of the forest. We forget that trees are alive and can regulate their leaf temperature physiologically. According to Ian Woodward in a recent *Nature* paper, the leaf temperature of trees in sunlight is self-regulated at close to 21°C; this temperature appears to be optimal for photosynthesis and is independent of the tree's geographic location, occurring in Arctic as well as in tropical regions. The leaf temperature is regulated by evapotranspiration. I have observed in the southern English summer that dark conifer-tree leaves maintain a surface temperature more than 40°C cooler than an inert surface of the same colour. On the scale of a forest as large as Amazonia or the boreal forests of Siberia, this has a huge effect on regional climate. Richard Betts and his colleagues at the Hadley Centre have pioneered the investigation into leaf temperature and its effects on both climate and the carbon cycle. As the leaves stay close to their optimal temperature physiologically the absorbed radiant energy of sunlight is mainly transformed into the latent heat of evaporation. It takes nearly 600 calories to evaporate a gram of water and meteorologists call the warmth stored this way 'insensible heat'. What some atmospheric physicists seem unaware of is the link between climate and the physiology of the ecosystem of the forest. When the strong feedbacks implicit in this link are included, especially the way a vast forest can melt away like the floating polar ice, it affects global as well as local climate. Feedbacks on the large regional scale may lead to punctuations of the kind illustrated in Figure 3.

Another illustration of the way that models are not of the real world concerns the water vapour in the air. On a cold morning at dawn we often see mist – fine water droplets floating like a grounded cloud in low places. The air where there is mist is close to fully saturated with water, 100 per cent relative humidity. As the sun rises and warms the air the mist fades and by early afternoon under a cloudless sky the relative humidity may be as low as 30 to 40 per cent. Large climate models have to assume that relative humidity is conserved, otherwise they become unstable. But in the real world relative humidity can be a truly important climate variable. The size of aerosol particles directly and rapidly changes with a change of relative humidity and so does their reflectance of sunlight and consequently the amount of heat that reaches the Earth.

The average surface temperature of the whole Earth is well constrained: the sun's output of heat is remarkably constant over the timescale of a hundred years and varies no more than 0.2 per cent, which is equivalent to a temperature variation of about 0.2°C. Just now, at the end of a long solar sunspot minimum, the sun's heat should be near its nadir. The Earth's orbit and its inclination to the plane of the solar system will also change little in the next hundred years. But, as we now know, quite small changes in the composition of the air or the nature of the land surfaces can have large effects. If the Earth ever did become a white reflecting snowball its surface temperature would be −24°C, exceptionally cold compared with now; but it has also existed for long periods in the past with even the polar regions tropical in temperature. Only 14,000 years ago we were in an ice age where glaciation at times extended as far south as the Alps in Europe and what is now St Louis in North America. It seems that the Earth can exist for long periods at a wide range of different climate states. The hot and cool stable states are well-founded facts of history, and we can explain them with a fair degree of confidence. What we do not know much about are the details of the move, say, from an ice age to an interglacial like now. This move seems to have been started by a small increase in heat received from the sun caused by a small alteration of the Earth's inclination and orbit, but there must have been substantial amplification through positive feedback to make it happen quickly. It is the similarity between the sudden changes then and what we are now doing that makes present-day forecasting so fallible.

Climate models based on atmospheric physics have their own peculiar dogma: almost all of them forecast a smooth, steady rise in temperature as carbon dioxide abundance increases. They seem to assume that nothing in the next thirty years will alter the course of global warming because our changes to the Earth's land surface and emissions so far have committed the system to warm about 2°C and its response time is slow. This is the basis of the IPCC recommendation to reduce emissions by 60 per cent by 2050 to avoid 'dangerous' climate change. The IPCC is right to think that it will take thousands of years to undo the harm that we have done and that in our terms there is no going back. They are also right about carbon dioxide emissions: the response time of the Earth to carbon dioxide change is of the order of 100 years. But it is wrong to think that nothing can happen rapidly in climate change. Aerosols in the atmosphere, snow and ice albedo, ecosystem response, and of course human response – any of these can cause a perceptible climate change within months. If the many apparently separate positive and negative feedbacks on climate synchronize coherently then the whole Earth system could heat or cool rapidly by as much as 5°C. I find it extraordinary that, given the depth of our ignorance, scientists are willing to put their names to predictions of climates up to fifty years from now and let them become the basis of policy. Surely they are not predictions, just speculations to assuage the fear of the dark clouds that loom on the climate horizon.

It is not mere speculation to challenge the idea that nothing can happen in the next thirty years that will alter the course of climate change. In a way the Earth has done the experiment for us, for when the volcano Pinatubo erupted in 1991 it injected enough aerosols into the upper atmosphere to cool the climate significantly for the three years that followed the eruption. It would be quite wrong to imagine that we can refine and improve those vast battle-ship models until they provide a clear and accurate view of the future climate. Even if we could, large volcanic eruptions are as yet un-predictable and could upset the forecast by injecting a vast cloud of cooling particles into the upper atmosphere. So could the use of geo-engineering to achieve under more control the consequences of these volcanoes. Numerous other natural events, admittedly less likely, such as the impact of an object more than 1 kilometre in diameter falling from space, or a repeat of the Maunder minimum when the sun's radiation declined a fraction of a per cent for one hundred years, human disasters such as a pandemic or a technological catastrophe of the kind predicted in Lord Rees's book *The Final Century*: all of these and others unknown may affect the climate and make distant prediction formidably difficult.

In addition to these uncertainties climate forecasters are obliged to model atmospheric physics when they should be modelling Gaia, or at least the whole Earth system of which the climate is one property. Administrators of science often imagine that a team composed of first-rate biologists, chemists and atmospheric physicists all working together, as in the IPCC, will solve the climate problem. In practice this may be no more likely to succeed than it would have been to seek the cause and cure for typhoid fever in Victorian times by analysing the fluctuating body temperature of the victims and then asking a team of biologists, chemists and physicists for the answer.

At this point I feel that a more general observation about climate change is needed. If we stand back and consider all the other perturbations possible to our self-regulating Earth, we see that the presence of 7 billion people aiming for first-world comforts is too much. It is clearly incompatible with the homeostasis of climate but also with chemistry, biological diversity and the economy of the system. Instability in any of these other properties of the Earth is potentially as disturbing as climate change and interacts with it. The acidification of the oceans by excess carbon dioxide is a single example of this multiplex pathology caused by an excess of affluent humans.

By assuming that the climate is mainly a physical property of the Earth's surface environment we leave out the important consideration of living organisms, including humans and their dependent species of crops and livestock, as an integral and interactive part of the climate system. This is the fundamental error of most computer climate models. It is an understandable error because the geophysics of the climate alone is beyond our present capacity, so that it seems absurd to consider including the even more complex biosphere as well. Naturally, science thinks that it has reduced the problem by subdividing it, and that presumably is why we have the mainly biological Millennium Assessment Ecosystem Commission, separate from the IPCC.

It would be wrong of me to suggest that climate modellers ignore the importance of the contribution of life on Earth to climate change. Climate modellers at the Hadley Centre and the University of East Anglia in the UK, at the National Centre for Atmospheric Research and other labs in the USA, and at Potsdam in Germany have all made or are making comprehensive dynamic climate models that do include the biota. I am familiar with the substantial contributions of Peter Cox, Chris Jones and Richard Betts of the Hadley Centre, of Tim Lenton, Andrew Watson and Peter Liss of the University of East Anglia, and of John Schellnhuber, Wernher Von Bloh and Stefan Rahmstorf of the Potsdam Institute for Climate Impact Research. But I think they would all agree that their work is far from complete. Then there are the climatologists Ann Henderson-Sellers, Kendal McGuffie and Robert Dickinson, who have tirelessly and often against strong opposition extended the competence of climate modelling by recognizing the need to include the biota in a dynamic role. For those interested in the arcane topic of climate research the book *A Climate Modelling Primer* by McGuffie and Henderson-Sellers is wonderfully rewarding.

We have not learnt from history. Before we became concerned with climate change, scientists and world governments were deeply concerned with the depletion of stratospheric ozone by CFCs. During that crisis there was near total acceptance of model predictions. Scientists were so convinced of the truth of their models that they rejected observations by Earth-orbiting satellites that saw the hole in the ozone layer over Antarctica. It took human observers – Joseph Farman, Brian Gardiner and Jonathan Shanklin of the British Antarctic Survey – to convince scientists that there really was massive ozone depletion and that the models were wrong. (They were in Antarctica at the time and were viewing the ozone layer with a Dobson spectrophotometer.) The reliance on models continues, and as I described earlier in this chapter, the vast hole that appeared in 2007 in the floating ice of the North Polar Ocean was not predicted to occur when it did. According to model predictions the meltdown was not expected before 2050. The real Earth responds to our actions in a way very different from that forecast by well-behaved models. These models also suggest a smooth rise of temperature as the abundance of carbon dioxide increases, and hint that temperatures can be lowered again by merely reducing the abundance of carbon dioxide. It seems that governments find forecasts like this easier to follow and more comfortable than the vacillating fluctuations of real observations. Large models have an authority from which policies can be drafted, conclusions drawn and grand pronouncements made at places such as Kyoto and Bali. But, despite the dedication of climate scientists equipped with the latest computer hard- and software, models are like ideologies, and possess a similar certainty, so that it is easy to forget that they are about abstract not real worlds. The modellers, except when corralled into consensus, express their forecasts properly in probabilistic terms.

A last reason for my disquiet about forecasts based on models arises because I earn my living and fund my research on the Earth as an independent scientist by selling inventions and advice. I have lived this way for nearly forty-five years and realize that it resembles the lifestyle of an old-time medical practitioner whose practice was in a small but prosperous town. This independent role has made me an observer, not just of the atmosphere, ocean and land surfaces but also of many of the human divisions of power and knowledge. These have included leading energy and chemical companies, and governmental agencies in Europe, the USA and Japan. I have also worked at many universities, including the United Nations University in Tokyo, and the intelligence agencies that have their own power to reveal the unexpected. For most of the time I was not much more than a wasp that had flown in through an open window, large enough to be noticed but not greatly affecting the conduct of business. Towards the end of the last century I was President of the Marine Biological Association (MBA) at a time when its laboratory in Plymouth was struggling for a degree of independence. For a few years I joined with my colleagues there in battle with a government that seemed driven, not always wisely, to control and centralize. (If you are curious to know more about this side of my life it is in my autobiography *Homage to Gaia*.)

This third component of my knowledge base has taught me that above all humans hate any conspicuous change in their daily way of life and view of the future. As Bertrand Russell put it, 'The average man would rather face death or torture than think.' The overwhelming wish to continue with business as usual applies far beyond the marketplace and may be a consequence of the cognitive dissonance I wrote about earlier. Business as usual is unfortunately how most of science is done even though we know that it has no place in science's probabilistic world. For practical and administrative reasons we cannot suddenly change the direction of research of a large and expensive laboratory built around a costly assembly of instruments, computers and specialized staff; this may be part of the reason why our forecasts do not agree well with expectations drawn from the history of the Earth.

These then are my reasons for thinking that climate forecasts decades ahead are at present too unreliable for planning detailed action. The task of the IPCC has barely started and their failure to account for even the current climate suggests they might need a new scientific approach, perhaps one that models the Earth as a single physiological system, not as a consensus model cooked from a biodiverse stew of scientific disciplines. Of course we should do our best to cut back damaging land use such as forest clearance anywhere and the farming of biofuels and cautiously prepare to reduce emissions. Until we know for certain how to cure global heating, our greatest efforts should go into adaptation, to preparing those parts of the Earth least likely to be affected by adverse climate change as the safe havens for a civilized humanity. In choosing havens safe from serious climate change we will need the guidance of the IPCC and perhaps they should be tasked to do this. Most importantly, we have to stop pretending that there is any possible way back to that lush, comfortable and beautiful Earth we left behind sometime in the twentieth century. The further we go along the path of business as usual the more we are lost.

The most important question in climate change is: how much and how fast is the Earth heating? I repeat that there is a trustworthy indicator of the Earth's heat balance, and that is the sea level. Its rise is a general and reliable indicator that cuts through arguments as to whether some glaciers are melting and others advancing and whether extra snowfall balances extra meltwater. The sea level rises for two reasons only: from ice on land that melts and from the expansion of the ocean as it warms. It is like the liquid in a thermometer: as the Earth warms the sea level rises. It is true that the level could suddenly increase if a large glacier in Greenland or Antarctica slipped into the sea, but this is most unlikely to happen unnoticed and its effect is easily discounted.

I sense the onset in science of a battle between those who live by theory and those of us who go out on to the Earth to observe and measure. The observers are the Cinderellas of science and always have been. Charles Darwin did not travel the Earth to prove a theory. He was a supreme observer and naturalist: the theory was developed later, some of it after he had died. The ocean is truly an *aqua incognita* and vitally important for the climate because it stores most of the extra heat of global warming. It is right to build theories of the ocean even though we know so little about it, but quite wrong to use them to make policies. First they must be tested by long-term observation and measurement, and that I think should be our first priority.

3

Consequences and Survival

When someone discovers, too late, that they suffer from a serious and probably incurable disease and that they may have no more than six months to live, their first response is shock and then, in denial, they angrily try any cure on offer, or go to practitioners of alternative medicine. Finally, if wise, they reach a state of calm acceptance. They know that death need not be feared and that no one escapes it. If the disease is cancer that wonderful organization, the hospice movement, founded by that true saint, Dame Cicely Saunders, often makes the end more seemly than the beginning. Scientists, who recognize the truth about the Earth's condition, advise their governments of its deadly seriousness in the manner of a physician. We are now seeing the responses. First was denial at all levels, then the desperate search for a cure. Just as we as individuals try alternative medicine, our governments have many offers from alternative business and their lobbies of sustainable ways to 'save the planet', and from some green hospice there may come the anodyne of hope.

Should you doubt that this grim prospect is real, let me remind you of the forces now taking the Earth to the hothouse: these include the increasing abundance of greenhouse gases from industry and agriculture, including gases from natural ecosystems damaged by global heating in the Arctic and the tropics. The vast ocean ecosystems that used to pump down carbon dioxide can no longer do so because the ocean turns to desert as it warms and grows more acidic; then there is the extra absorption of the sun's radiant heat as white reflecting snow melts and is replaced by dark ground or ocean. Each separate increase adds heat, and together they amplify the warming that we cause. The power of this combination and the inability of the Earth

46

now to resist it is what forces me to see the efforts made to stabilize carbon dioxide and temperature as no better than planetary alternative medicine.

So far as I am aware no one at Bali or earlier UN meetings was directly concerned about Gaia or considered the response of the living Earth to what we are doing to it. In fact, as the Earth warms, and long before the dateline of 2050, the output of greenhouse gases and the albedo changes caused by the Earth itself may exceed the total warming effect from all the extra gases we have added. The assumption that the climate can be stabilized by a reduction in emissions at a carbon dioxide abundance of 550 ppm and a global temperature 2°C higher than normal has no secure foundation in science. Instead the Earth system could already be committed to irreversible change, even if we implement in full the recommended 60 per cent reduction of emissions.

It is surprising that politicians could have been so unwise as to agree on policies many decades ahead. Perhaps there were voices from scientists who warned of the absurdity of such planning, but if so they do not seem to have been heard. Even if we cut emissions by 60 per cent to 12 gigatons a year it wouldn't be enough. I have mentioned several times before that breathing is a potent source of carbon dioxide, but did you know that the exhalations of breath and other gaseous emissions by the nearly 7 billion people on Earth, their pets and their livestock are responsible for 23 per cent of all greenhouse gas emissions? If you add on the fossil fuel burnt in the total activity of growing, gathering, selling and serving food, all of this adds up to about half of all carbon dioxide emissions. Think of farm machinery, the transport of food from the farms and the transport of fertilizer, pesticides and the fuel used in their manufacture; the road building and maintenance; supermarket operation and the packaging industry; to say nothing of the energy used in cooking, refrigerating and serving food. As if this were not enough, think of how farmland fails to serve Gaia as the forests it replaced did. If, just by living with our pets and livestock, we are responsible for nearly half the emissions of carbon dioxide, I do not see how the 60 per cent reduction can be achieved without a great loss of life. Like it or not, we are the problem – and as a part of the Earth system, not as something

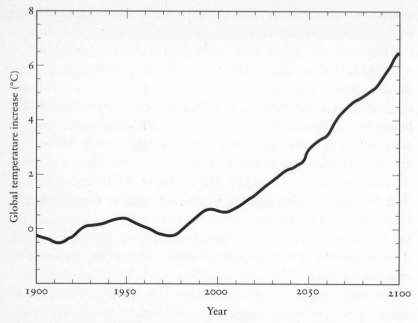

Figure 4. *The forecast of heating for the northern hemisphere for the coming century according to the model predictions quoted by Peter Stott in his 2006 article on the unusually hot European summer of 2003. The line shown was hand-drawn and should not be taken as more accurate than a blackboard sketch.*

separate from and above it. When world leaders ask us to follow them to the inviting green pastures ahead they should first check that it really is grass on solid ground and not moss covering a quagmire.

The only near certain conclusion we can draw from the changing climate and people's response to it is that there is little time left in which to act. Therefore my plea is that adaptation is made at least equal in importance to policy-driven attempts to reduce emissions. We cannot continue to assume that because there is no way gently to reduce our numbers it is sufficient merely to improve our carbon footprints. Too many also think only of the profit to be made from carbon trading. It is not the carbon footprint alone that harms the Earth; the people's footprint is larger and more deadly.

We already face the adverse consequences of a total accumulation of greenhouse gases amounting to over 430 ppm of carbon dioxide equivalent – the loss of land-based ecosystems, the desertification of the land and ocean surfaces, and the loss of polar ice; these act together in positive feedback and probably commit the Earth to irreversible heating. There may be no alternative but the direct use of the global cooling techniques discussed in Chapter 5 on geoengineering, including an attempt to massively decarbonize the atmosphere by burying charcoal. Whether or not these efforts succeed in cooling the Earth to its previous self-regulating interglacial state, we have to prepare for failure by adaptation.

The crux of it is that there are far too many of us living as we do – Paul and Ann Ehrlich said so forty years ago in their book *The Population Bomb*. But we did not listen. They tended to exaggerate, but their insight about the dangers of overpopulation was right. In theory, we could eat less and save energy, but in practice we never will, unless made to do so. The consequences of our overgrowth and its emissions are only slightly different from those caused by the photosynthesizers (single-cell plants) who also grew and multiplied 2 billion or so years ago, and so changed their world that whole ranges of anaerobic ecosystems were condemned to an underground existence. Their pollution was oxygen, a poisonous, carcinogenic and fire-raising gas that life, including us, has evolved to benefit from. Like the photosynthesizers, we could not have avoided reaching our current overpopulated and unsustainable state. We are what we are and there is little that we could have done to avoid what now seem adverse changes; we should not feel guilty about it.

If our leaders were all great and powerful they could ban the keeping of pets and livestock, make a vegetarian diet compulsory, and fund a huge programme of food synthesis by the chemical and biochemical industries: doing this might limit the loss of life to pets and livestock only. It is encouraging that the chairman of the IPCC, Dr Pachauri, has recommended a vegetarian diet as a way to go. Almost certainly it will never happen this way, and people will continue with farming, business and government as usual. Changes in lifestyle, agriculture and eating habits are not a popular political option and governments are more likely to take the easy way of using tax and subsidy to drive farms, industry and the public in whatever direction their political ideology favours. We often forget that an industrialist's duty is to his or her company's shareholders, not to the community or the government, and certainly not to the planet. Industrialists are not greedier or more insensitive than the rest of us, but tax and subsidy distort their ability to make a profit and so they will usually choose inefficient but profitable sources of energy and agricultural products over long-term sensible and efficient but less profitable choices. This is why industry will back renewables, carbon trading and biofuels, which are neither efficient nor sensible, but are immediately profitable. Nuclear energy is profitable, even without subsidy, but, as with the purchase of a house through a mortgage, the profit is deferred. In the present economic climate distorted by subsidy, nuclear is less attractive to industry.

Meanwhile climate change progresses remorselessly, now driven by feedbacks from the Earth as well as our increases in emissions and land use. There is no tipping point; we are sliding down a bumpy slope that grows ever steeper to the future hot world. Even in the survival havens where climate change is gentle enough to allow the continued growth of food there will be disasters and difficulties. Thus in the more fertile parts of Europe unaffected by heat and drought, including the Netherlands, the UK and Ireland, rising sea level and storms may lead to catastrophic inundations. Much of London is likely to be flooded, and the underground transport system disabled. The Netherlands may be made uninhabitable. Even temporary flooding with salt water greatly lowers the productivity of farmland. When considering food and energy supplies we have to keep in mind that the immediate needs of human consumers is only one part of the problem. We need also to sustain the infrastructure of the cities, the housing, health and other services, including schools, waste disposal and transport. Too easily forgotten is Gaia's need: we have to leave enough natural ecosystems on land and in the ocean for planetary self-regulation.

Climate change is fickle. Events in early 2008 have made many in Europe and the USA doubt that global heating was proceeding on schedule or was our prime cause for concern. The climate did not yet seem bad enough for urgent action and our minds were filled with fears of a recession or depression of the monetary climate. Indeed a planetary physician who was looking at the bedside chart of our supposedly ailing planet would note that despite global heating, which is confirmed by the persistent rise of sea level during the past ten years, the average global temperature has not changed appreciably over the same period; some good climate scientists even think that there may have been a small fall in temperature during the present century. It is true there have been scary symptoms such as the extraordinary melt of Arctic ice in the summer of 2007, but despite these worries the Earth's fever did not seem to be worsening. Moreover, 2008 was turning out to be a cool wet summer in north-western Europe and parts of the US, not at all what was expected of global heating. This apparent remission of the Earth's illness was reflected in Nigel Lawson's thoughtful brief *An Appeal to Reason*. His book reads like a breath of fresh air coming from an open window in an overheated conference room. Most climate-change deniers fail to hide a vested interest in the status quo and are unconvincing or even boring. But here is a book denying global heating written with passion but still a proper detachment, as if the author were the defence counsel for the deniers of climate change.

I think that he is right to criticize the hype that goes with the public response to global heating. But I wholly disagree with his denial, and think there is only the smallest possibility that the world will not grow hotter as I have described in Chapter 2. For now it is useful to compare the Earth with an iced drink. You will have noticed that the drink stays cold until the last of the ice melts, and so to some extent is it with the Earth. A great deal of the heat of global heating has gone into warming that huge lump of water, the ocean, and into melting ice. This may be one of several reasons why it has not warmed more, but once the ice has melted and the mixing of the ocean waters has reached a dynamic equilibrium, global heating will proceed even faster than before. Lawson's book forces us to think about the Earth and what we are doing to it in the larger context. I applaud his astringency and his disapproval of the trendy populism that now attaches to anything and everything seen as green.

Human nature, the behaviour that comes from the intelligence that evolution has given us, impairs our chances. We are like high-flying raptors – hawks and eagles that evolved to catch their prey from the skies and do it exceedingly well. But what would be the fate of eagles if all their prey moved to live underground? They are not adapted to fly in tunnels or caves and their fine eyesight would not be useful in the dark. We are perfectly evolved to live as hunter-gatherers. The wings of our brains are fine-tuned by evolution to survive in the world of 1 million years ago, but we are as ill-equipped to survive on the 21st-century Earth that we have made as is a hawk in a cave. Our intelligence is not something transcendental but a property that evolved to fit us into our niche, like the tough beak of a woodpecker evolved to fit it into its world where the food supply is tree-bark bugs.

Our contemporary industrial civilization is hopelessly unfitted to survive on an overpopulated and under-resourced planet, deluded by the thought that clever inventions and progress will provide the shoehorn that fits us into our imaginary niche. I think it is better if we accept and understand how poor is the chance of our personal survival, but take hope from the fact that our species is unusually tough, has survived seven major climate catastrophes in the last million years, and is unlikely to go extinct in the coming climate catastrophe. Geneticists, interested in the evolution of humans, have observed that at one time in the last million years we passed through a genetic bottleneck in which our ancestors might have been as few as 2,000. Gaia, fortunately, is much tougher and as a living planet has survived for over a quarter the age of the cosmos.

A specific example of an evolved mental attribute once favourable for survival but now a serious handicap was given by Michael Shermer in his column in the *Scientific American* of August 2008. He took a recent medical controversy to explain why thinking anecdotally comes naturally but thinking scientifically does not. The controversy is about whether or not autism is connected with the administration of vaccines to children. On one side are some parents who notice that shortly after vaccination autistic symptoms appear; on the other are scientists who find no causal link between vaccination, or the preservatives in the vaccine, and the symptoms of autism. The anecdotal associations, especially when they are amplified as a story by the media, are so powerful that they cause people to ignore contrary scientific evidence. Shermer goes on to say that the reason for this cognitive disconnect is that we evolved brains that pay attention to anecdotes because false positives (believing that there is a connection between A and B when there is not) are usually harmless, whereas false negatives (believing that there is no connection between A and B when there is) may take you out of the gene pool. Our brains are belief engines that employ associative learning to seek and find patterns. Superstition and belief in magic are millions of years old, whereas science, with its clever ways of circumventing false positives, is only a few hundred years old.

How very similar is the autism vaccine connection to the anecdotal belief that there are clusters of leukaemia victims in the populace around nuclear power stations. I know as a scientist that this is nonsense, but try convincing a woman who lost a relative who happened to live in the vicinity of a nuclear installation that the likelihood is vanishingly small. This is why it is so easy to persuade the gullible multitude that the harmless mobile phone you use, or the nearby power cable, is a danger.

If our present world is wholly unsustainable, how do we retreat from it sustainably? To answer this question it is useful to think of a nuclear submarine as a microcosm of the Earth. It has to exist beneath the waves for periods as long as half a year and always sustain an environment healthy for the submariners. The energy comes from stable and reliable nuclear reactors. The reactor in the submarine is so well shielded that submariners are exposed to less radiation than anyone else alive. At 100 metres deep no cosmic or Earthly radiation penetrates; but for us on the surface, in addition to the cosmic radiation and the natural radioactive elements in the soil and walls of our buildings, our reactor, the sun, is unshielded except for the thin layer of air that separates us from space. We soon discover how poorly we are protected when we sit in the summer sun too long and suffer a radiation burn. The air of the submarine is as well regulated as is the atmosphere we breathe at the surface, and the water supply recycled and reliable. In the manual of the officer who does the regulation is a warning never to allow the oxygen abundance to rise above 21 per cent: this is not in the interests of the sailors' health alone but because the fire risk nearly doubles for every extra per cent of oxygen in the air, and fire in a submarine is deadly.

The carbon dioxide also needs regulating for the crew continuously exhale it and too much would make breathing difficult. Just imagine how catastrophic it would be for submariners if fossil-fuel energy was used inside the submarine. Like the submariners, we enjoy constant and reliable energy from our great nuclear reactor in the sky, and Gaia regulates the air and water supply for us. No one would doubt that the submarine is limited as to the population it could sustain so why should we imagine that the Earth has an unlimited capacity for people?

Our troubles now are eerily similar to those of the unfortunate crew of a sunken submarine too deep for rescue but on which there are a few escape capsules in which some can float up safely to the ocean surface. How like the Earth, now so overpopulated that only a small portion will reach the remaining habitable land. I think that is how we are, but bad as it is it gives the chance for our species to survive.

The greatest harms global heating causes are not the dramatic surprises of unprecedented weather events such as violent storms and floods of rainwater or almost unbearable heat. Harm comes from prolonged and unremitting drought. According to the forecasts (IPCC report from Working Group II, 2007) many parts of the world will experience such a lack of water by 2030. Saharan conditions will extend into southern Europe, as they are experienced in Australia and Africa. Heavy rain will fall but when it is hotter than the mid 20s Celsius it does little good. Increasing heat and forest ecosystem destruction to provide farmland will continue and hasten the conversion of rainforest to scrub and desert. Provided there is ample energy, the heat can be withstood personally using air-conditioning technology; indeed, conditions in cities of the hot regions are unlikely to be much worse than those in Baghdad, Alice Springs or Phoenix now. Death comes from drought when neither food from crop plants nor water is available.

When we look at projected future climates we see that much of the continental areas will become barren because of drought. This will have appalling consequences for the already overcrowded nations like China, India and parts of Africa. Life on Earth is wholly dependent on water and three-quarters of the bulk of nearly all life forms is water. Without its abundant supply food crops will not grow, and to irrigate all the land now naturally watered by rain is an impossible task. There will be much smaller areas where this is done and these too, like the older civilizations along the Nile and the Euphrates, will be havens.

Emphatically this does not mean there is nothing we can do. We do not have to sit and wait for rescue like those unfortunates who remained, as instructed, in their offices in the Twin Towers on the 11th of September 2001. We can move where it is safe. Of course we will procrastinate and continue to deny global heating and try to fool ourselves that there is no urgent need to move, but the increasing rate at which polar ice melts, ocean levels rise and the climate zones migrate warns us that the Earth system is already moving and so soon must we. Moreover the observed positive feedback on heating makes it unlikely to slow or stop before the next stable state is reached. Through geoengineering we may ameliorate some of the early consequences of heating but I greatly doubt that we have the wisdom or intelligence to reverse it. Like the skier who accidentally starts an avalanche, there is little we can do to stop its destructive course.

So are all our efforts to become carbon neutral, to put on sandals and a hair shirt and follow the green puritans pointless? Can we go back to business as usual for a while and be happy while it lasts? We could – but not for long. Apart from a lucky break of a natural or a geoengineered kind, in a few decades the Earth could cease to be the habitat of 7 billion humans; it will save itself as it dispatches all but a few of those who now live in what will become the barren regions. Those who leave for the cooler, still fertile regions have a better chance of surviving, and if enough of us are saved this way it could benefit Gaia as well. It seems that enough of us could survive to carry on our species, but there is an overriding need that reduces the carrying capacity of the Earth even further and that is the requirements of Gaia herself. There is much more to survival than human needs alone. To sustain the climate and composition of the Earth, Gaia needs the ecosystems, the forests and other vegetation on land, and the algae of the oceans to sustain life. Otherwise, our planet would move inexorably to the intolerably hot and utterly barren equilibrium state, something that eventually would be an average between Mars and Venus.

What is certain is that it is our duty to survive. Our greatest efforts therefore should go to learning how to live as well as is feasible on the soon-to-be-diminished hot Earth. We in Britain live on one of the safe havens where life can continue in the heat age. In certain ways we are like the passengers aboard a ship that has diverted to take on board refugees escaping from some drought-stricken land. To the refugees we are their lifeboat, but the captain and officers of the ship have to decide how many we can take – who can be allowed to board and who must remain and take their chances? Fairness suggests a lottery, but common sense rules out so simple a selection. The sick, the lame and the old would have to stay behind and take their chances along with passengers who felt called to help them. On ships it used to be women and children first, but some men would be needed – what would be the right ratio of the sexes? I suspect that it would not be far from equal, for that is the proportion that natural selection has chosen.

There is no simple number for the carrying capacity of the Earth for people. It depends on the way the people live. Are they at one extreme vegans, or at the other carnivores? Do they farm and so displace natural ecosystems? Are they industrialized, and what is the impact of their industries? In addition to these human properties the Earth itself is not a constant. The number it can carry varies with its state. If mainly desert, the number will be small; if well watered and rich in nutrients, it can be as many per square mile as in Bangladesh. Were we hunters, carnivorous top predators, it is unlikely that even a fertile Earth could carry more than 10 million of us. As gatherers, especially if vegan, it could be 100 million or more. With science and technology present, the numbers are imponderable, and we have proved that 7 billion is possible for a short period. But how many will be in balance on an Earth 4°C hotter than now? It might be as little as 100 million if the carrying capacity of the land surface of a hot Earth falls to 10 per cent of what we have now. All that we can usefully say about the carrying capacity of the Earth for humans is that it changes rapidly.

One natural thing in our favour is that more than 70 per cent of the Earth is ocean and that proportion will grow as the sea level rises, otherwise the Earth would heat even faster; but unfortunately a warm ocean is much less productive than a cool one. We must concentrate on the encouragement of ocean ecosystems mainly with planetary climate regulation in mind, although some food and fuel would be by-products. Looking again at the world two or three decades ahead, we see that the largest areas of land possible for habitation are in the north and south temperate and Arctic regions – Patagonia and southern Chile, Canada, Siberia, Alaska, Northern Europe including Scandinavia and the western oceanic coasts – in addition to islands such as the British Isles, New Zealand, Tasmania, and many smaller ones dotted across the oceans. We do not yet know whether they will be too dry and hot for crop growth: the geological record of the last hot period 55 million years ago suggests that the Arctic basin was tropical in temperature and that vegetation was abundant. The rest of the continents will not be wholly barren: there will be oases and river courses still watered well enough for plants to grow.

Any attempt by the surviving civilization to continue to farm land that should remain as natural forest, or to burn fossil fuel, could lead to disaster but the temptation would be great, for the Arctic is the site of vast quantities of oil, gas and coal. If we used these as we use them now, we might become our own executioners and cause the death of Gaia as well. The Earth would then be left hot and barren with no life other than a few thermophiles: too sparse a biosphere for a self-regulating planet.

For the present I am assuming that global temperature will rise during this century at least as severely as the mid range of the IPCC predicts, and that its direct physical consequence, sea level rise, will proceed as it has been doing since 1990, nearly twice as fast as predicted. Nothing is certain; and I have to allow that none of this may happen. Instead, one or more of the several proposals to geoengineer the Earth and stop global heating might work, or some natural event such as a series of giant volcanic eruptions might intervene, or the models that predict the climate are even more wrong than I thought they were. The best contradiction of all would be to discover that the idea described by Johannes Lehmann in *Nature* in 2007 allowed us to take a massive quantity of carbon dioxide from the air by making charcoal and then burying it in the soil. I have described it more fully in Chapter 5 and I do think it has a chance of halting global heating. But having said this, and knowing our obstinate desire to continue with business as usual, I doubt that if it is tried, it will be done to a sufficient extent to realize its promise. Our good intentions are too often forgotten like the unread promises we acknowledge with the click of a mouse at the end of a long and unreadable legal statement that appears on the screen with every computer application we buy. I will write on assuming that you have clicked the button and agreed.

To understand a little of what lies before us I shall focus on the islands where I live because they provide a history and an example of human response to a threat which, although far less severe than global heating, was sufficient to make survival an imperative. For these islands this was the Second World War in 1939, and it was certainly enough of a threat to stir the response now needed. Let me tell you how I personally experienced the onset of it, when I was twenty years old.

The path ran along the edge of fields recently harvested for their crop of grain; it went between Chelsfield and Orpington, some fourteen miles south-east of London's centre. As I walked in September 1939 the London suburbs already encroached upon the countryside. The fields had a tired look, as if they were about to give up the game and retire for good beneath a permanent crop of semi-detached houses planted by their new owners, the developers. But my angst about the ruin of rural Kent was rudely disturbed when suddenly and to my amazement the air filled with the sound of air-raid sirens. I walked on wondering if soon the sky would fill with bombers, but instead the sirens sounded the all-clear. And so the Second World War started with a false alarm; indeed in terms of war nothing much happened on mainland Britain for another nine months. There seems to be a close parallel between the events and the feelings we had then and those now. I was not quite that archetype, the man in the street or on the Clapham omnibus, but was close enough: a young man on a footpath, fairly sure that real war would soon begin even though there were still deniers, among them experts and politicians.

Seventy years later events in far places, such as the melting of the

Arctic ice, the collapse of glaciers in Antarctica, the droughts and famines across Africa, and the occasional extra-fierce tropical storm give us now that same anxiety that the war in Spain and the incursion into Bohemia gave in the 1930s. We somehow sense that it will be our turn soon but we continue our business and pleasure as usual and perhaps put a solar heating unit on the roof, just as we dug air-raid shelters in our gardens back then. In 1938 my father, although retired and in his late sixties, dug an impressive shelter that went forty feet below the garden with a concrete chamber under the house and entrances on both sides of the house. He started digging in 1938 and finished the job before the war began. How odd that when war threatens, individuals instinctively prepare for the worst, often with futile acts, while our elected representatives and the civil servants who help them prepare instead for the previous war by building battleships and constructing fortresses like the Maginot Line.

Gaia, like God, helps those who help themselves. It was not enough in 1939 to dig personal air-raid shelters, nor is it now enough to genuflect with small green gestures; nor to put windmills and solar panels on the roof to supplement the electricity supply; nor to hold meetings before that great religious symbol of spin, the giant white wind turbine and sing hymns about salvation for the planet. Not only must we survive but we must stay civilized and not degenerate into mob rule where gang leaders promote themselves as warlords. For this we have to take effective local action now. Most of all, we have to secure our supplies of food and clothing and, if we continue city life, energy. These islands, although among the few areas of the world least threatened by global heating, are at the same time among the least well supplied with food and energy. We have grown so used to an ample supply of food from abroad that we forget that in the Second World War, when food imports were scarce, we nearly starved. We have indigenous sources of fuel but they are fast declining. The land available for agriculture competes with housing and industry, and unless we act soon more of it may be disabled as the numbers inhabiting our small nation steadily increase.

Just as in 1939 we had to give up on a massive scale the comfortable lifestyle of peacetime, so soon we may feel rich with only a quarter of what we consume now. If we do it right and with enthusiasm it will not seem a depressing phase of denial but instead, as in 1940, a chance to redeem ourselves. For the young, life will be full of opportunities to serve, to create, and they will have a purpose for living. It will be much tougher for the old, but as that still viable wartime comedy *Dad's Army* revealed, far from dull. Whatever happens, it will be quite a change from the banalities of city life now.

Those were my memories of the UK nearly seventy years ago. Other participants in the Second World War, such as Germany and Russia, would not provide evidence as benign, because for them the cycle of victories and defeats profoundly affected the extent to which they were in control of their fate. The occupied nations of Europe certainly endured stress and privation, but they were in no way their own masters and do not make a good comparison between then and now. The USA was of course deeply involved, but the shortages and incidents on the mainland were small and of shorter duration. Probably the closest in experience to the UK was Japan, and it would be interesting to have a similar comparison of life now and in their war.

So let us try to imagine what life might be like for an ordinary family living in the town of Reading, about thirty miles from London, in 2030. Assume that the models of the IPCC predict the course of events as shown in Figure 4, but allow that they may underestimate. The forecast global temperature rise is 1.8°C and the sea level rise, 12 cm. Our family will hardly notice any change, especially since they have had twenty years in which to adapt. In war there are long quiescent spells then sudden violence and panic, and so it may be with climate change. The Thames will have flooded seriously on a few occasions from excessive rainfall, but so far the sea will not yet have reclaimed the Thames Valley as a tidal creek. Perhaps new housing will still be appearing on the flood plain, in between the floods – new housing will be needed now that the population has risen to perhaps 80 million, as refugees from Europe and the world come in. The most noticed things will be the dullness and shortages, and the expense of food and energy. If Europe has failed to abandon its love affair with renewable energy and we have failed to build adequate supplies of nuclear energy, electricity will be ruinously costly and brown- and blackouts will be endemic. The family will whinge and grumble but somehow muddle on. But much of the rest of the world will be changing to scrub and desert and (as our Government's chief scientist, John Beddington, has recently warned) drought and famine will be taking over the once-fertile Earth. Closer to home, just across the Channel, summer heat will have grown unbearable, despite the widespread use of air-conditioning. Food production will be falling as drought and heat make growth ever more difficult. Elaborate schemes to irrigate using desalinated sea water will alleviate some of the loss, but at a huge price in energy. The flow of climate refugees will continue, with many settling in huge encampments possibly near the ethnically similar communities of earlier immigrants.

Assume that this is approximately a true picture of the course of events if we let them just happen. But what if at some time in the next few years we realized as we did in 1939 that democracy had temporarily to be suspended and we had to accept a disciplined regime that saw the UK as a legitimate but limited safe haven for civilization. It could be forced upon us by a weather event like that of 1953 when a storm tide in the North Sea devastated parts of the Thames Estuary and the Netherlands. In all, hundreds died. A similar event now could devastate much of the Netherlands, London and its hinterland. Perhaps this would be enough to bring to the fore some Churchill whose rhetoric would fire the nation to make the effort needed to adapt properly to change instead of just patching its problems in an incoherent way. Orderly survival requires an unusual degree of human understanding and leadership and may require, as in war, the suspension of democratic government for the duration of the survival emergency. Good leadership is vital and I know that Sir Crispin Tickell has already proven his capacity to inspire a UK government to steer itself in the right direction when global heating first became an issue in the 1980s. He has continued with this task and inspired leaders of other nations, and I hope that he can do it again for our survival.

I suspect that effective action to sustain this island community will come from some form of internal tribal coherence and rare leader-ship, not from international or European good intentions. With luck the same will apply with the other havens. There will be time enough for internationalism during the stability of the long hot age. We have no option but to make the best of national cohesion and accept that war and warlords are part of it. For island havens an effective defence force will be as important as our own immune systems. Like it or not we may have to increase the size of and spending on our armed forces. Perhaps the next generation of scientists and engineers will be competent and serve the Earth as general practitioners serve us in medicine. In wartime old dogs were quite quickly taught new tricks. The first truly great environmental disasters will usurp the political agenda and displace many false ideas hampering change. As in war there could be the rapid application of new technology to climate and survival problems. I hope that it will work, but I do not think humans as a species are yet clever enough to handle the coming environmental crisis and I fear they will spend their efforts trying to combat global heating instead of trying to adapt and survive in the new hot world. So let us prove Garrett Hardin wrong when gloomily he said in 1968 that our condition is truly tragic; for in tragedy there is no escape. We can prove him wrong by surviving.

Because I am old I often think of Gaia as if she were an old lady of about my age. I can already hear Pecksniffian colleagues complaining, 'You are doing it again – anthropomorphizing the Earth, talking of it as alive.' But I say to them, 'If it is not alive then how can it die?' And die she will when the sun's heat becomes more than can be withstood. Those of us who have thought about it see her lifespan extending no more than 500 million years from now. It sounds a lot, but since she is now 3.5 billion years old she has already lived nearly 88 per cent of her life. If I can reach 100 then, intriguingly, at eighty-nine as I write, I am now the same relative age as Gaia.

The end for me could easily be the catastrophic consequences of a disease like influenza that meant no more than a few days off work when I was young. So it is with Gaia: for her, cosmic missiles in the form of asteroids or comets continuously threaten. The last to hit was 65 million years ago and did devastating damage. Should one of these hit as she approaches 4 billion years old, that and the extra heat from the sun may be more than she can stand and the great system that has kept mainstream life on Earth for more than a quarter the age of the universe will end. Of course, like my corpse when I die, cells and bacteria will live on for a while, but the dead planet will be unable to sustain an environment fit for life.

I am glad that I have no notion of my own end; so with Gaia all that can be said now is that elderly planets like elderly people are liable to die from maladies that the young and vigorous can shrug off. Our obligation as an intelligent species is to survive; and if we can evolve to become an integrated intelligence within Gaia, then together we could survive longer.

4

Energy and Food Sources

Two important changes in our future energy supplies emerged after I started writing this book about a year ago. What has changed is the public perception of nuclear energy and the recognition that solar thermal energy is the most promising of the 'renewable' energy options. It even seems possible that by using both of these we can significantly reduce our dependence on fossil fuels, although the greater part of the energy we use will still be drawn by burning fossil fuel for at least a decade from now. This is inevitable for practical and economic reasons because it takes at least a decade, even with willing support, to replace an energy source on a global scale, and there is unlikely to be a noticeable decline in fossil fuel use before these two alternatives are well established. Not only do power stations take time to replace, but long-distance high-power transmission cables need to be installed for solar thermal energy coming into Europe from southern Spain or the Sahara. In the USA, where there are large areas of sunlit desert in the southern states, solar thermal energy is even more promising. Some features of fossil fuel use are also difficult to replace, for example liquid and low-pressure gaseous fuels deliver huge energy flows in a very short time. When you draw 10 gallons of fuel into the tank of your car, in the sixty seconds it takes the energy flow is equivalent to the full output of a 25 megawatt power station. No conceivable battery or super-capacitor could be charged this rapidly, nor can we yet envisage a battery-powered long-distance passenger aircraft. It is true that liquid fuel can be synthesized from carbon dioxide using nuclear or solar electricity, but this must wait until the energy sources themselves are well established.

When comparing energy sources it is usually assumed that the energy is used for electricity production. This ignores the sizeable use of fossil-fuel energy by industry and for warmth in winter. At present the major energy sources are: fossil-fuel combustion, nuclear energy and energy from flowing water. None of the fashionable 'renewable energy' sources have yet made a significant impact on supply, and of these only solar energy has a chance of delivering in time to offset climate change. Wave and tide energy have considerable promise but are unlikely to deliver in the next two decades. In some regions, such as the Midwestern plains of the USA, wind power might become a minor but significant source, as it might in places where the trade winds blow constantly.

At present the intermittency of supply hampers wind energy, but this would be less of an objection if the power was used for making fresh water from the sea, for irrigation, and for pumping water from inundated land. In Europe there is a rush to large off-shore wind energy installations: it is difficult to see how these could ever produce electricity reliably and economically. I suspect that the rush to wind energy is driven more by ideologically based subsidies than by good sense, and if the imminent economic recession is prolonged they will be an unwanted burden. On our small, overcrowded islands wind energy is not a sensible choice at all, because of the huge spaces needed to gather even a small fluctuating quantity of electricity; every acre of land may soon be needed for food production and for re-creation. For the developing world simple solar voltaic cells can gather sunlight by day, store it as electricity in a rechargeable battery, and provide power for lighting and communications: such low cost installations enrich the lives of those who live in tropical lands. Tidal energy is well proven but limited to suitable locations, such as the Severn Estuary in the UK. The most sensible statements on the proper use of available energy systems were in a 2008 report by Professor Ian Fells and Candida Whitmill: they rightly observed that some forms of 'renewable' energy are appropriate as a supplement for the UK's energy needs, and reminded us that a tidal energy scheme for the Severn Estuary could supply 5 per cent of the UK's electricity.

In theory it is easy to improve our use of energy by the avoidance of waste. In practice this is hard to do in times of plenty and there is not yet the will to do it. A great deal could be done to lessen the burning of fossil fuel for energy by simple technological improvement, and I hope that our present diminished economy will encourage it. Why, for example, has it taken so long for manufacturers of light sources to move from the exceedingly inefficient hot filament lamps to light emitting diodes (LEDs), or for car manufacturers to switch from fuel-hungry monsters to small but adequate cars that are three or more times less fuel-devouring. It is easy to blame the manufacturers, but we who buy and use their products share at least half of the blame. As always, Europe and North America assume that the problem and its solution lies with them alone, but in reality the developing world and the new consumers of India, China and soon South America are beginning to dominate the production and use of energy. In a similar and equally deluded way we assume that the human presence on the Earth is all that matters, yet when considering energy and our use of it we must never forget that the natural flux of energy and those essential gases oxygen and carbon dioxide from the biosphere is nearly twenty times larger than all of our emissions, and it changes as the world warms.

We are bemused by carbon, and when we talk and think about our abuse of the Earth we concentrate almost exclusively on greenhouse-gas emissions from transport and industry and from domestic heating and air-conditioning. We try to convince ourselves that if we sufficiently improved our carbon footprint all would be well again and business as usual could continue. In reality increasing numbers of people increases the population of livestock and of the area of land we use for ourselves. True enough, the world total of domestic and industrial emissions of 30 gigatons of carbon dioxide annually is far too great, but so are the consequences of too many people competing for land with the natural forests of the world.

SOLAR ENERGY

The sun delivers 1.35 kilowatts of energy to every square metre of the Earth it shines upon directly. An area of desert land in the southwest of the United States of 10,000 square kilometres (3,600 square miles) receives enough sunlight and heat to power steam generators that could supply all the electricity needs of the USA. A similar idea has been considered by the EU using the Sahara or even southern Spain as a location for solar thermal power stations.

It is important to distinguish between the two main methods of converting solar radiant energy to electricity – solar voltaic and solar thermal. Sunlight can be directly converted to electricity by solar voltaic cells – semiconductors that absorb light and deliver an electric current to wires connected to the cell. This is a practical way of providing electricity to space vehicles and satellites and is well established. The material usually used is silicon, either as the expensive single crystalline form of the element or as the cheaper but less efficient amorphous silicon. The conversion efficiency for the production of electricity from sunlight varies between 10 and 20 per cent. In recent years more efficient alternatives to silicon have been tried. They are usually based on some exotic and rare elements, especially gallium, indium, selenium, tellurium, arsenic and cadmium. The highest efficiency reached is nearly 30 per cent, but only under laboratory conditions. Most recently semiconductors based on polymeric carbon compounds have been tried. So far solar voltaic cells are not serious candidates in economic terms for large baseload installations. But they have promise of further development and are of immense value in small-scale uses, such as for charging the batteries of portable devices.

With solar thermal energy, sunlight falling uninterruptedly on permanent desert is gathered as heat radiation and used to generate electricity. In one method, first developed in Australia, an array of long flat mirrors with grooved surfaces is focused on to a long pipe suspended above them that serves as the boiler and super-heater. The steam made is used directly to drive conventional steam turbines. The absence of solar energy at night is offset by the use of steam accumulators – a nineteenth-century invention – which can store enough energy to keep the turbines running for hours after the sun goes down. (The details of these solar-energy power stations were described in *Scientific American* in 2008.) Recent technological developments in cables capable of carrying huge quantities of electricity over thousands of miles mean that power stations can be sited at greater distances from their customers. As an example, the energy output of two nuclear power stations now flows from France to the UK along cables that lie on the floor of the English Channel. These cables can operate at near a million volt potential and use direct current instead of the alternating current carried by the pylons of present-day supplies. The new main cables cost about $1 million a mile, not a great deal in the context of supplying energy on a continental scale. Unlike many renewable energy proposals, this energy source is not visionary. Its main components already exist and have been tried – prototype solar thermal plants are running in Arizona and the calculations look good; let us hope that this is the next practical large-scale energy source that we can use. Much depends on how soon it could be implemented: if it takes twenty years or more, it could be too late. Meanwhile declining fossil-fuel energy and emerging nuclear fission energy can fill the gap.

NUCLEAR ENERGY

We should regard nuclear energy as something that could be available from new power stations in five years and could see us through the troubled times ahead when the climate changes and there are shortages of food and fuel and major demographic changes. Those in Britain should think of the troubled years of the 1970s and early 1980s, when industrial conflict over coal threatened electricity supplies. It was the availability of nearly 30 per cent of the electricity we used from nuclear energy that sustained the nation and prevented the quarrel turning into a civil war. The only thing that prevents an immediate build of new nuclear electricity is legislation put in place by previous governments and unreasoning fear.

There are now over 442 nuclear power stations in the world and together they produce 17 per cent of all the electricity used, about the same percentage as hydroelectricity. Other sources of renewable energy – biofuels, wind, etc. – produce only 2 per cent. The safety record, their cost and the local acceptability of these fission-powered stations make them the most desirable of all sources. So why in the first world do we still persist in the falsehood that they are uniquely dangerous?

I think we fail to welcome nuclear energy as the one good and reliable power source because we have been grievously misled by a concatenation of lies. Falsehood has built on falsehood and is mindlessly repeated by the media until belief in the essential evil of all things nuclear is part of an instinctive response. Here are some of these untruths and their refutations:

Nuclear energy emits large quantities of carbon dioxide and is therefore as polluting as burning fossil fuel. This is nonsense: a nuclear power station while running emits no carbon dioxide at all. A small amount is emitted by the transport that delivers the fuel and takes away waste, and this is listed in Table 1.

Nuclear	4
Wind	8
Large-scale hydro	8
Energy crops	17
Geothermal	79
Solar	133
Gas	430
Diesel	772
Oil	828
Coal	955

Table 1. *UK Government's Energy Technology Support Unit report on the output of carbon dioxide in grams per kWh of energy produced by power sources.*

Nuclear is hardly dirty compared with the rest, and the figure given includes all emissions over its entire operation, taking into account mining and processing the ore and decommissioning and waste disposal. The figure for wind does not include the large (375 g/kWh) output of carbon dioxide from the back-up power station needed when the wind is too high or too low.

It is often said that nuclear waste is uniquely deadly and will persist for millions of years and poison the global environment. All pollution by chemical elements persists. Lead pollution from a mine, smelter or the factory where it is made into things for commerce lasts for ever; and the same is true of mercury, arsenic, cadmium and thallium: these toxic elements are permanently with us. What is remarkable about nuclear waste is that it fades away. In 600 years the high-level waste from a nuclear power station is no more radioactive or dangerous than the uranium ore from which it originated. Far more importantly, there is hardly any nuclear waste to worry about. The yearly output of waste from a 1000 MW nuclear power station is enough to fill a London taxi. Now perhaps you see why I would welcome its burial at my home in Devon. It would be a useful source of heat.

Even Government committees such as the Committee on Radioactive Waste Management (CoRWM) propagate nuclear falsehood: one of its representatives said that there is enough nuclear waste in Britain to fill the Albert Hall five times over. In fact, after 40 years of generating nuclear energy, there is barely enough to fill one Albert Hall. Compare this with the mile-high mountain, twelve miles in base circumference, of solidified carbon dioxide that the world makes every year. The nuclear waste is a minor burial problem but the carbon dioxide waste will kill us all if we go on emitting it.

Opponents of nuclear say there is a shortage of uranium. This is utter nonsense. Uranium is not a rare element and even if it cost as much as gold this would barely affect the price of electricity produced in a modern power station. Britain has a great deal of pure uranium and plutonium ready to produce energy but because of the Blair administration's anti-nuclear folly this valuable fuel is marked to be decommissioned at great cost.

Another falsehood is the claim that the emissions from nuclear installations are a threat to life and health. Nuclear radiation is natural and a normal part of our environment: we and all life evolved with it. The total emissions from the UK nuclear industry are 500 times less than that of the radon gas we breathe every day of our lives. Radon comes from the rocks and soil and is natural.

These errors would be harmless were they not continuously propagated and amplified by all branches of the media. Much of it is unconscious. Here is an example from a piece by a favourite columnist of mine, Matthew Parris, writing in the *Spectator* recently. He had this to say: 'Nuclear Power. A method of generating electricity which, like road traffic accidents (which killed 3,000 in Britain last year), poses an appreciable risk to life and limb; but instead of doing so through small, frequent accidents, threatens big but rare dollops of danger. There have been none recently; the issue has faded in the public mind.' The juxtaposition of 3,000 deaths per year from road traffic accidents with nuclear accidents, unintentionally suggests comparability. The truth is that deaths from accidents having a nuclear cause in the UK nuclear industry in 2007 were zero and in the fifty years of its operation here also zero. True, there was a nasty accident at Chernobyl twenty years ago that killed a total of seventy-five, mostly brave firemen and rescue workers. Then there was the expensive industrial accident at the Three Mile Island power station where no one was even hurt but which frightened a whole nation. In fifty years in the whole worldwide nuclear industry no more than one hundred have died. Compare this with the tens of thousands who died in the coal and oil industries and the hundreds of thousands who died in making renewable energy or from the consequences of using it. Yes, hydroelectricity is 'renewable' but who notices how dangerous it is until a dam bursts?

Few engage in propaganda merely as a hobby, so who benefits from it? The media does to some extent but generally, I repeat, innocently. The main benefactors of anti-nuclear propaganda are much more sinister, but let's consider the media first.

As a scientist who lately became a writer I realize just how difficult it must be to write a good column weekly for a year. Most of us could manage it for a few weeks, but a year demands an intelligent imagination and stamina that few possess. The column writer must be the pampered aristocrat of the media compared with the lumpen mass of writers who deliver the flow of news. I have often wondered about the life expectancy of journalists; those I met seemed to work under conditions that implied an unusually stressful life. I could not conceive a harder task than having to interview the hostile and unwilling and then type up the experience to a deadline while obeying political and editorial constraints; small wonder that even good journalists rate an entertaining easy-to-tell story higher than dull and uninspiring truth. Good stories require a cast of demons and angels; in war, whether cold or hot, figures like Stalin or Hitler can easily be cast as arch-demons and we can play the role of angels. In the looming global environmental war the truth is that we all are the demons, and this is not an acceptable role for us; still less is it acceptable to story-writers. We have had to invent new angels and demons.

The green lobbies and political parties owe their existence to the unending flow of good stories about environmental disasters. *Silent Spring* was a globally effective scare story that is often said to have started the modern green movement. So, on a local scale, was the 1970s television series *The Good Life*, where we the viewers could identify with the angelic good efforts of an ordinary couple to be 'green'. The baddies were the usual suspects: oil polluting the sea and killing birds; coal that had to be dug by overworked and underpaid miners, both the products of malign multinational companies moved by nothing other than profit. We always forgot or ignored that for most of the time the coal industry was nationalized so we were the owners of this polluting industry as well as the users who generated the pollution. In the 1970s and early 1980s it was the black carbon fuel that was seen as evil, not the invisible product carbon dioxide. We also ignored the dangers of carbon dioxide then because the minds of scientists were fully occupied by the threat to stratospheric ozone posed by the CFCs.

At this time nuclear energy was cheaply, safely and securely making 30 per cent of the electricity we used in Britain and doing it quietly in the background. It took a medium-sized industrial accident in a Soviet nuclear power station at Chernobyl in the Ukraine to trigger the monstrous anti-nuclear energy story that has haunted the world ever since. The accident, a steam explosion, happened in an unstable reactor that was undergoing an unwise and improperly planned experiment. The whole sad event was a sequence of false steps that could only have happened under the corrupt statist politics of the Soviet Union. The seventy-five people killed were almost all either plant workers or those called in by the state to clean up the mess. It was a trivial event compared with the industrial disaster in the city of Bhopal in India where, in the early hours of the morning of 3 December 1984, an accident at a pesticide plant released forty tons of methyl isocyanate gas into the night air. The cloud drifted over the town and killed 3,800 people instantly and many times more in the weeks that followed. The Bhopal disaster is usually cited as the world's worst industrial disaster, but how often do the media mention it compared with the much lesser disaster at Chernobyl?

At Chernobyl the smoke from the burning reactor mingled with the air mass moving westwards towards Europe. Radioactivity can be measured with instruments of extraordinary sensitivity, so it was not surprising that the cloud was detected as far away as the UK. Had the wind blown to the east we probably would never have put the word Chernobyl in our vocabularies; as it was, the media amplified the story until governments fearful of accusations of inactivity did incredibly foolish things. The Swedes performed a massive cull of reindeer and we in the UK banned the sale of lamb from the hills of Wales and Scotland. The ban and the cull were justified on the grounds that the meat was dangerously contaminated when, in fact, it was only a few times more radioactive than meat is naturally, and insignificantly dangerous to health. The Italians closed their nuclear power stations and the Germans proposed to do the same. Even the BBC for years after solemnly announced that there would be tens if not hundreds of thousands of radiation deaths across Europe. Repeated investigations by teams of physicians from United Nations agencies have failed to find any evidence to support these gloomy predictions. Radiation scientists who could have challenged this nonsense chose to keep quiet. When considering the consequences of Chernobyl it is useful to recall that the radiation exposure to all of us who were alive in 1962 from nuclear bomb tests was 100 times greater, and despite this we now live longer than ever.

Let me give you another example of the false fears spread by anecdotal evidence. My wife Sandy and I live in a remote part of England and our telephone wire crosses farmers' fields on poles for six kilometres; often in storms it is damaged and we can no longer communicate or send e-mails this way. In practice, our BlackBerry mobile telephone keeps us always in touch. What madness it would be for us to reject the chance to communicate because we feared cancer from the microwave radiation of mobile telephones. But this is what more than half of us do nationally by rejecting nuclear energy on the same insubstantial grounds.

I am not alone in concern over the reticence of scientists to reject the falsity of these anti-nuclear allegations, and their unwillingness to engage in the larger cause, our survival. In a remarkable essay entitled 'Scientific reticence and sea level rise', published in *Environmental Research Letters* in 2007, James Hansen made the same comment that reticence among scientists was inhibiting them from warning of the threat of a potentially large sea-level rise. We scientists are too introspective and seem to have reached a state where sometimes a neat theory supported by computer calculations trumps observation and experiment. We seem less inclined to test our ideas in the real world and we no longer seek the judgement of nature, preferring the judgement of our peers. Like our theological forebears, we are starting to produce truth in a virtual world instead of discovering it.

The first believable and effective environmental angels in the UK were the Campaign for Nuclear Disarmament (CND) and Greenpeace. The appalling consequences of Hiroshima and Nagasaki made it wholly natural for them to cast nuclear weapons as quintessentially demonic, and CND had the authority to do this directly from their ecclesiastical leader Monsignor Bruce Kent. The aim of CND and Greenpeace was to make the UK first, and then the world, a nuclear-free zone. It was great good fortune for these campaigners to receive the gift of prominent nuclear weapon test explosions by the Americans and the French. We are all to an extent tribal nationalists; consequently even liberals in Britain thought that there must be something good in a movement that was simultaneously both anti-American and anti-French. There were of course Soviet nuclear weapon tests, but the Soviets wisely kept them secret even from their own inhabitants.

And so was founded the idea that everything nuclear, including the use of nuclear energy to make electricity, was bad, unhealthy and sinful. It became easy for the media to use it in their stories. By this means, and at no cost to CND and its fellow travellers, their message was amplified until now no political party in the UK has the courage fully to endorse nuclear energy as the greenest, cheapest, safest and most secure source of electricity. Just as bad, the more sensible members of these anti-nuclear lobbies who would like to retract their objection to nuclear energy are unable or afraid to do so.

So who benefits from the propaganda? Normally the media can smell a rat better than a hungry terrier, and I was slightly surprised that they did not wonder more about the murder of the Russian dissident Litvinenko in 2007 in London. He was cruelly poisoned by a few hundred nanograms of the radioactive element polonium 210. It is an element of the sulphur family of the periodic table which is readily incorporated into the simple biochemicals that feed the cells of our bodies so that when swallowed it soon finds its way to every cell of the body. The poisonous dose of polonium was carefully chosen. It was enough to cause about seven polonium atoms to disintegrate in every cell of the victim during the few days of life left to him. Polonium emits helium atoms travelling at a fraction of the speed of light and they plough through the vital structures of a cell so that one or two of them is usually sufficient to kill a cell. An evil way to kill someone: a slow, unstoppable, tortured death. There is ample evidence that the agents of the murder were Russian and the container of the radioactive element was leaky enough to leave a trail from the airliner that brought the assassin to London to the hotel where the poison was added to a cup of the victim's tea.

What an opportunity was missed by some imaginative journalist or thriller writer to have set a scene somewhere in Moscow with a cast of professionals from security agencies or energy corporations. The meeting would have included the assassin, toxicologists, and managerial bureaucrats. Someone says, 'You realize, do you, that a poisonous dose of polonium 210 will cost about ten million dollars? Why not use ricin – we know that that's a reliable poison and a lot less visible to the media; moreover, it will cost less than a dollar.'

Another bureaucrat adds, 'Yes, and to make the polonium we have to seek time on a reactor which is already fully occupied with other important tasks.'

At which a senior manager intervenes. 'Gentlemen,' he says, 'the purpose of this action is not merely to punish a traitor – and that alone needs visibility and media amplification – but more importantly to keep the West frightened of all things nuclear. Our future as a world power depends on our ability to make them wholly dependent on us for their supply of oil and gas; their use of nuclear energy would free them of this dependency and we could lose our ability to make the world go the way we wish. Ten million dollars is nothing in that cause.'

This scene is no more than a figment of my imagination, but it could have made a good story at the time. Moreover it grows more credible as we move into the twenty-first century, when political power and business opportunity will more and more be linked to energy supply. It would be naive to expect energy companies to stand aside and see their profitability hampered by inexpensive nuclear energy, and the same must be true for the thwarting of national aspirations. The cash flow of nuclear industry is tiny compared with that of oil, gas or coal companies, and the money available for advertising the advantages of nuclear is proportionately less.

Fear of nuclear has become so deeply entrenched that if an engineer in a Japanese nuclear power station drops a wrench on his foot and needs first aid it is given headline exposure in our news-papers as a 'Serious accident in Japanese nuclear power station'. The death of a hundred or more Chinese miners in an underground coal mine explosion rates not more than a small paragraph in the depths of the same paper.

What I have just written is no exaggeration. In July 2007 an earthquake in Japan shook a nuclear power station enough to cause its automatic shutdown; the quake was of sufficient severity – over six on the Richter scale – to cause significant structural damage in an average town. The only 'nuclear' consequence was the fall of a barrel from a stack of low-level waste that allowed the leak of about 90,000 becquerels of radioactivity. This made headline front-page news in Australia, where it was said that the leak posed a radiation threat to the Sea of Japan. The truth is that 90,000 becquerels is just twice the amount of natural radioactivity, mostly in the form of potassium, that you and I carry in our bodies. In other words, if we accept this hysterical conclusion, two swimmers in the Sea of Japan would make a radiation threat.

FOSSIL FUELS

This book is not the place to list comprehensive statistics of fuel reserves and production. With modest effort and discernment they are available on the internet, but I will include some comments on the fuels now in use and on our food prospects for the coming century.

There are huge reserves of coal, oil and natural gas still left, and in addition there are even larger reserves of less efficient fuels such as tar sands, shale and peat. The catch is this natural fuel is renewed very slowly, at one-hundredth the pace we burn it. It is not the amount that we have burnt but the rapidity with which we are doing it that matters. However, if in your car you used petrol or diesel fuel made from atmospheric carbon dioxide using nuclear energy or solar thermal energy you would be doing something good for the planet. It is not the fuel that is damaging but the balance sheet of its production and use. It all depends, as my friend and mentor Chris Rapley, Director of the Science Museum, often reminds me, on population. If there were only 100 million of us on the Earth we could do almost anything we liked without harm. At 7 billion I doubt if anything sustainable is possible or will significantly reduce fossil-fuel combustion; by significantly I mean enough to halt global heating. Seven billion living as we do, and aspire to do, is too many for a planet that tries to self-regulate its climate.

If we could collect all of the carbon dioxide emitted from the burning of fossil fuel before it reached the air, we might be able to bury it deep underground in spaces from which it could not escape. Attempts to do this are under way around the world, but these must surely fail to stop global heating because their total effect can never be large enough. More details of these and other schemes for resisting climate change are in Chapter 5.

Crude oil was first exploited commercially in Pennsylvania in the USA in 1850. Soon after its discovery it was realized that the distilled product of the crude material was more saleable. Petroleum made its first mark on the world as kerosene, or as it is called in the United Kingdom, paraffin. It was used worldwide mainly for lamps and cooking stoves. In some parts of the world, especially Africa, it is still so used. In the 1920s and 1930s, I was often sent to remote farms for school holidays. Quite often the cooking was done on a paraffin stove and the food tainted by the malodorous fuel. When travel by jet aircraft became commonplace it made the smell of paraffin familiar. One unremarked advance in petroleum chemistry has been the removal of most of the intensely odorous components of paraffin and diesel fuel so that now the smell is almost bearable.

By the end of the nineteenth century the distillation of crude oil had developed hugely. The products were light gases; volatile liquids (for example petrol/gasoline); less volatile liquids (paraffin and diesel fuel); heavier fractions that were the fuel oils for ships and furnaces; the almost non-volatile lubricating oil fraction; and finally what was left of the distillation process was tar or asphalt. I have spent some of my life in or close to oil refineries in the UK and the USA. For me there was always an element of awestruck fascination in a visit to one of these orderly entanglements of pipes amid the deafening roar of the refineries operating in the background. At first I visited as a pioneer of gas chromatography, and showed their chemists how to separate and identify the huge range of hydrocarbons that go to make up a fuel like petrol. Something of great interest to them, since the quality of the fuel – its octane value – depended on its composition. In turn I learnt that running a refinery was rather like piloting a plane that almost never lands. The crude oil arriving in tankers from distant places has to be pumped to quite limited storage tanks from which it then flows continuously to be distilled. The seven or more separate streams from the distillation column also flow continuously through a series of further refinements until, without stopping, they enter the product storage area. Liquid fuel is somewhat like electricity: once made it has to flow until burnt or used. The possibility of a long strike of tanker drivers was a permanent anxiety for refinery pilots; the unplanned stopping of a refinery was somewhat like landing a plane on a motorway.

All fossil fuel is biological in origin, and this is even true of natural gas, which is one of the main energy providers for much of the world. Not long ago it was considered the cleanest, most efficient and least polluting of all fossil fuels, and much of Europe's reduction of carbon dioxide emissions comes from its use of readily available natural gas. The molecule of methane contains four hydrogen atoms for its one carbon atom, so when burnt roughly half the energy comes from the hydrogen content whose combustion product is water. The carbon, of course, gives carbon dioxide, but only half as much as coal or oil for the same amount of energy.

Not only this, methane can be burnt directly in thermodynamically efficient gas turbine systems, from which the waste heat can be used for town heating. Woking, in Surrey, has one of the lowest pollution figures of any town in the UK largely through the use of combined heat and power generation. If the world had stayed as it was in 1960, conversion to methane burning might have been enough by itself to meet the need to halt climate change. We have become so profligate in energy use that pressure on the world's reserves of methane has led to price rises that still further taint the once pristine quality of the gaseous fuel. It is rarely mentioned that if methane leaks into the air before it is burnt, it has a greenhouse effect about twenty times greater than carbon dioxide.

Coal is the truly dirty fuel. From the marital discord caused by a miner who leaves his carbon footprint on his wife's new carpet to the London disaster of 1952 when more than 5,000 died from coal-smoke poisoning in one night, it has always had this reputation. Countless numbers still die and are made ill by coal smoke in the world, especially in China and Mongolia. Yet it is not the coal itself that kills but the inefficient way we burn it in open fires. London's air, apart from traffic pollution, is now almost clean to breathe, although 33 per cent of all the electricity we use still comes from burning coal. In 2008 Sandy and I were invited to have breakfast at a London hotel with James Rogers of Duke Energy, and Mary, his wife. My friend Stewart Brand had made the introduction and thought that we would both benefit from the meeting: he was right. I found Jim Rogers, a leading figure in the huge American coal indus-try, to be as concerned with our future as I was, and wonderfully practical. We shared the opinion that there was neither the time nor the resources to bury the carbon dioxide effluent of coal-fired power stations on a global scale. It was invaluable to hear from him how huge the usage of coal worldwide as the prime energy source was. Whether or not we can reverse the climate change now happening depends on how fast we can do it. The meeting confirmed my view that globally there is very little chance of going back to the world of one hundred years ago. Jeff Goodell in his book *Big Coal* gives the best and most recent account of our huge problem with coal.

RENEWABLE ENERGY

Renewable energy is usually defined as energy generated from natural resources – such as wind, sunlight, flowing water, tides, geothermal heat, biofuels and the burning of biomass – which are naturally renewed. It sounds good for both humans and the Earth, and many truly believe its full use could 'save the planet'.

Renewable energy is something that comes from Gaia whereas fossil fuel and nuclear energy are man-made and therefore dirty. This is wholly untrue and a myth that goes back at least to Rousseau. We imagine that somehow we can all live naturally and that natural energy and organic food are fundamentally different and better in quality than what is manufactured. It may seem that blowing wind and flowing water are more natural than a nuclear reactor, but they are not: there were natural nuclear reactors in what is now Gabon in Africa that kept warm the micro-organisms that made them 2 billion years ago. Also the turbines that extract renewable energy from the wind or water flow have to be manufactured: they wear out and need renewing, and so do the steel tower and the blades that rotate. All the energy we use, except nuclear, is second- or third-hand solar energy, and the term 'renewable energy' makes no sense in our present world. So what makes it so attractive even to hard-headed businessmen? They are attracted by the subsidies offered by governments driven by the pressure of a fashionable and trendy green ideology. The same persuasive force makes them penalize what is perceived as ungreen: coal, oil and nuclear energy.

Can energy be distinguished as renewable or not according to its source? No, it cannot. The first law of thermodynamics, one of the three great laws of the universe, states 'Energy is always conserved', but there is nothing about it being renewable. In this universe energy cannot be renewed: all you can do is take it, use it and be grateful.

The word 'conserved' is easier to understand. Think of a pint glass of cold water: if you add to it a single teaspoon of boiling water and mix it in, the glass of water will become barely perceptibly warmer. But the total heat energy of the water in the glass will have been raised by exactly the amount of heat energy in the spoon of boiling water. Energy was therefore conserved. Interestingly, if you had an instrument-maker build you a tiny engine, the difference in temperature between that spoon of boiling water and the glass of cold water could have been used to drive a tiny dynamo and make electricity. But however you used the electricity, the energy would still have been conserved.

The adjective 'renewable' is used as a human value judgement: it has no basis in science. But because we are not gods and goddesses who can produce energy or matter from nothing, we have to obey the laws of the universe, and surprisingly this implies that anything we make is natural. A four-wheel drive SUV and the fuel in its tank is as natural as a termite nest. Without life on Earth neither of them could exist, nor could the car be driven; we too easily forget the fuel is useless without oxygen. SUVs and the fuel in their tanks are not intrinsically good or bad, although what is done with them can be. So what is all the fuss about? The fuss is that there are so many of us that we burn fuel more than 100 times faster than the Earth can renew it.

WIND ENERGY

Like nuclear, wind energy is one of the more contentious and vexatious of energy sources. Used sensibly, in locations where the fickle nature of the wind is no drawback, it is a valuable local resource, but Europe's massive use of wind as a supplement to base-load electricity will probably be remembered as one of the great follies of the twenty-first century – an example of impressive engineering misused by ideology and as inappropriate as passenger transport by hydrogen-filled airships.

I must declare a special personal dislike of large wind turbines onshore. My home is in the county of Devon in south-west England, which is one of the last areas of Britain to retain the small-scale, checkerboard pattern of fields enclosed by hedgerows that have made our countryside so admired around the world. The South-West Coast Path goes 630 miles from Minehead in Somerset to Poole in Dorset, and passes along the Devon coast north and south: this coast includes some of the most magnificent scenery of Europe and part of it has rightly been chosen as a World Heritage Site by UNESCO. I chose to live here because I treasure this, one of the few remaining areas of countryside largely unaffected by urban influence or industrial farming, and see it as an example of how people and the land can live together in a seemly way. To make this county the site of an industrial-scale source of electricity from wind energy is to me as philistine as placing a sewage farm in Hyde Park or Central Park, yet this is the intention of our government, albeit under heavy pressure from the EU. A senior civil servant of the Department of Trade and Industry, in the time when Patricia Hewitt was the Secretary of State, referred to those like me who objected as Nimbys (not in my back-yarders), a term of abuse for those who resisted the unassailable good intentions of her department. I am proud to be called a Nimby, for my backyard is the countryside and I see that land as the face of Gaia.

Were these wind farms truly efficient and capable of resolving our power needs, I might be persuaded to grit my teeth and endure their ugly intrusion, but in fact they are almost useless as a source of energy. It would take 1,000 square miles of countryside to provide enough land for a 1 gigawatt wind-energy source. The wind blows only 25 per cent of the time at the right speed to generate a useful quantity of electricity; therefore this monster would need the back-up of a near full-sized fossil-fuel power station to supply electricity whenever the wind blew too much or too little.

In addition to the negative propaganda directed at nuclear energy, there are almost as many untruths propagated about the favourable qualities of wind energy. Take for example the British intention to build the world's largest wind farm in the Thames Estuary, which would have 341 turbines occupying an area of 230 square kilometres. It is claimed to be a 1 gigawatt project and therefore equal in output to a typical nuclear power plant. In the hype attending it is the claim that it will provide enough electricity for one third of London's homes

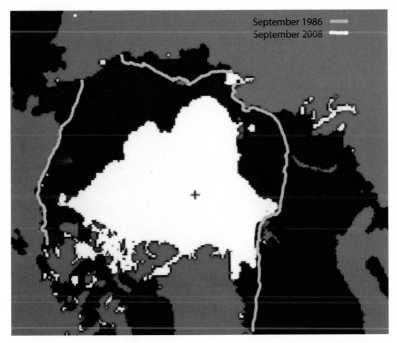

1. The 'Ice Hole', the vast area of Arctic floating ice that melted in the summers of 2007 and 2008. Like the 'Ozone Hole' it was unpredicted and is potentially deadly.

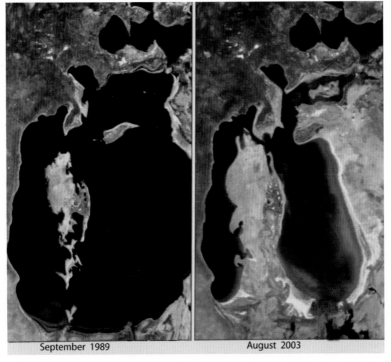

September 1989   August 2003

2. The vanishing of the Aral Sea, once the second-largest freshwater lake.

3. White Knight 2, the lift plane that carries the Virgin Galactic space plane to its launch at over ten miles up.

4. The small space ship from which I hope to see the face of Gaia.

5. The Devon countryside near my home, now under threat from industrial-scale wind farms.

6. A wind farm in Scotland, whose erection was encouraged by the Scottish Green Party.

7. The electron capture detector, ECD, accurately measured the CFCs, PCBs and pesticides in the world environment and so enabled the environmental movement.

8. The author standing above French high-level nuclear waste.

9. Drought in Australia, which may be usual elsewhere before long.

10. The Namibian desert, a landscape which again may be more widespread further down the line.

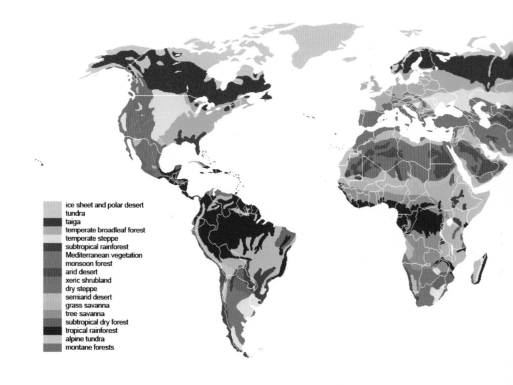

ice sheet and polar desert
tundra
taiga
temperate broadleaf forest
temperate steppe
subtropical rainforest
Mediterranean vegetation
monsoon forest
arid desert
xeric shrubland
dry steppe
semiarid desert
grass savanna
tree savanna
subtropical dry forest
tropical rainforest
alpine tundra
montane forests

12. World oceans showing vast areas of barren water, deep blue.

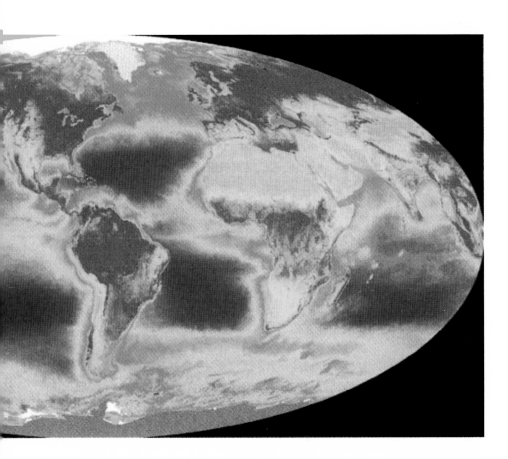

11. The spread of vegetation around the world.

13. Vegetation in the wild in Central Brazil. These tropical forests are a vital part of the Earth system.

and save the emission of 1.9 million tons of carbon dioxide. It sounds good until you realize that a full-sized, presumably coal-burning power station, emitting copious amounts of carbon dioxide, will have to be built to back it up when the wind does not blow. Its real averaged output would be only 400 MW of fluctuating electricity. If it were steady, which it is not, it would be enough for 830,000 homes each consuming 4,200 kWh yearly. I am glad the oil company Shell had the wisdom despite subsidies to pull out of this flawed project.

To survive on these islands with a population perhaps as large as 100 million requires a constant and reliable source of electricity from indigenous fuel. It would be madness to attempt it without nuclear energy. It is sad that so many of the green movement and their intellectual followers still oppose nuclear on grounds as insubstantial as a fear of hellfire and Satan.

Source	Footprint	Pollution	Subsidy needed
Coal and oil	2	10	No
Gas	1	5	No
Nuclear	2	1	No
Solar thermal	150	0	No
Solar voltaic	150	2	Yes
Wind	30,000	4	Yes

Table 2. *Illustrates for different energy sources their relative area cover (footprint), greenhouse-gas emissions, and whether or not subsidies are needed in order for them to be profitable.*

Let us compare the energy sources now proposed. The footprint of coal and nuclear power stations are comparable and a 1 gigawatt station occupies about 30 acres, or approximately six soccer pitches. Similar output gas-fired power stations occupy only about 15 acres. But solar power stations would require 5 square miles of desert at latitudes as low as 30° (so far distant from Britain). Wind energy requires 1,000 square miles for a 1 gigawatt onshore installation. The only entirely non-polluting source is solar thermal; wind and solar voltaic are in practice polluting because they would need back-up from fossil fuel when not producing. The disposal of the small quantity of nuclear waste I have considered as pollution even though it has no climatic significance.

Now that extravagance is no longer in fashion there is no excuse to maintain subsidies for any energy source. There is considerable choice, and inefficient sources such as wind in the UK should be abandoned in favour of nuclear energy. In Europe and the USA the promise of solar thermal energy increases the options.

Our greatest need in the UK will be a secure supply of food and energy. Soon the growing appetite of the world for both, and the worsening climate, will make the supply from abroad increasingly more expensive, and we will be driven more and more to produce food and generate electricity from our own indigenous resources. We in Britain are no longer a major engineering and manufacturing nation and may have to leave the engineering development of our energy supply to those nations better equipped to do so. The worst of all possibilities would be for us to become the test bed for unproven technology, and this is what is happening now with wind turbines. It takes a long time to turn even a good idea into a practical source of energy and we do not have the time. Let me explain.

Climate change may force us to develop new sources of food and energy; we need to react promptly and know the time it takes to establish something new on a global scale. The time will be determined by engineers and inventors, and their ability can be summed up in that phrase used by patent lawyers, 'reducing to practice'. It sounds easy but is the crucial process that turns an inventor's rough sketch on the back of an envelope into something that might be handy to have in your kitchen. Good engineers, through a long series of small steps, can turn an amateurish idea into polished practical utility. This way the first steam engines were slowly improved until they became the reliable and crucial part of nineteenth- and early twentieth-century industrial civilization.

The gestation period from a seminal idea to globally marketable product is far longer than we usually think – indeed it is on average about forty years. It takes that long to improve the first rough working model into something that affects nearly everyone. The first suggestion of television came sometime close to 1900 but it was not until the 1950s and 1960s that television sets entered homes worldwide. The Wright brothers flew their first plane in 1900 but it was another forty years before we could begin to use the atmosphere as an alternative medium through which to travel. Those in love with technology who believe all the words written by imaginative science writers find it hard to believe that, despite Moore's Law by which computer chips double their speed every one and a half years, it still took forty years from the first working electronic computers in the Second World War to the widespread use of computers in the home.

It will probably take at least ten years to make real on a global scale abundant energy from solar thermal energy or tidal energy, but from the expensive daily advertisements of energy companies you might imagine that cheap and abundant green energy was on sale now. It is not: the attraction of green energy schemes is simply the feedback of a portion of the subsidy, not that they are inherently good or could compete in the marketplace on their merits alone. It is only rarely that encouragement by subsidy is efficient. Subsidizing air travel in the 1920s would not have made it competitive with rail travel or passenger ships. Air transport had to await a full understanding of aerodynamics and the development of jet engines before it truly took off with all of us on board.

It is equally foolish to imagine that a wartime approach, like the Manhattan Project, would rapidly speed up progress. In practice this kind of approach may work well for the development of an existing possibility, but it can hinder new invention. Babbage in the nineteenth century was a genius to do as well as he did in designing a mechanical computer, but no amount of subsidy then would have accelerated those essential steps, the invention of the transistor and later of integrated circuits, that led to your own personal computer. Rushing invention is only rarely successful. Even the Manhattan Project had to build on the pre-war discoveries of Hahn, Meitner, Pierls and other scientists; its end product was a bomb, albeit a very powerful one, and it still took forty years from the discovery of the neutron by Chadwick in 1932 before nuclear energy became a practical source of energy supply on a global scale. It was not until the 1970s that electricity from nuclear fission was a significant part of the supply to every home in the UK.

THE VANISHING FACE OF GAIA

The novelist C. P. Snow described the disagreement in Britain between humanists and scientists in his book *The Two Cultures*. From about 1980 the humanists have been victorious, and the credibility and character of scientists and engineers has been downgraded. Much of the failure to prepare for and counter global heating comes from the inability of otherwise able politicians and civil servants, who are only rarely scientists, to understand the message from good scientists and engineers, and distinguish between genuine and alternative science. Too often the strident over-optimistic cries of entrepreneurs or trade lobbies trump sound and practical advice. Global heating is already here and we cannot afford to wait any longer. We need to start preparing our defences now.

FOOD AND LIVING SPACE

If Britain is to become the haven for Europeans escaping global heat, as well as energy we need a secure supply of food, sufficient to keep as many as 100 million well fed. Just as important, we need housing for them and cities with a full infrastructure for shops, schools, hospitals and workplaces. We have been amazingly careless about land use, and fooled by globalization's cornucopia of food that we seem to think will go on for ever. Rising food prices, like the first gusts of a great storm, herald the famine soon due. Now that the days of plenty are ending, how can we use the acreage of Britain to grow the food we need?

The yield from modern agribusiness farms is far greater than that from the traditional farms I know and love, and on a crowded island we may have no option but intensive farming, at least on the land that is best farmed that way. A particularly productive area of the UK is East Anglia, but some of the best of it is low-lying and in places is already below sea level. It is vulnerable to a large storm-driven high tide, and rising sea levels increase the threat.

It will be hard for those with a love for the gentler countryside we once enjoyed, but there must be sensible give as well as take. Cities need parks and the nation needs National Parks and the land now sensibly used by the National Trust and similar organizations.

We are lucky to have long stretches of superb coastal scenery and well-established footpaths along it. The coast is no place for growing food, or for that matter for energy farms. As in wartime, farming will evolve to meet our need for food, and the countryside will look very different from now. Intensive growth in greenhouses and transparent plastic tunnels is already happening, and as imports from abroad decline our need will increase for market gardening, growing vegetables and fruit. I wonder if we will still have cattle and sheep – a very inefficient way of producing food – or whether waste-eating poultry and pigs will provide our meat.

My dream is that we will discover how to synthesize all the food we need from carbon dioxide, nitrogen, water and a few minerals. Given abundant energy, it is not difficult to synthesize simple amino acids and sugars, which are the normal food of living cells, from air and water. This could then be the feedstock of bulk animal or vegetable tissue that could be harvested as food. Perhaps a biopsy sample of the leg muscle of a single Aberdeen Angus bull would in this way provide steaks for a multitude. (Something similar to this kind of synthesized food already exists in a commercial form. It is a cultured myco-protein produce which supermarkets sell under the brand name 'Quorn'.) Then the footprint of food production would shrink until whole swathes of land could return to Gaia. Maybe we could redeem technology and return this way to the natural world that was there before we began to use fire. Massive crop failure in future adverse climates would give food synthesis an immediately vital role. But like other technological dreams we do not have the time to do it now.

First we have to decide and plan how to house the 100 million, for that determines the area of land we still have left for growing food. My inclination would be for the compact cities envisaged in Lord Rogers' book *Cities for a Small Planet*. High-density living is not only possible but far better than suburban and exurban sprawl. New towns of this kind would set free land instead of, as now, competing with it. In such tight cities nearly everything needed could be no more than a walk away. Pedestrianized areas in towns and cities are already showing the way to go. As an addicted walker I dream of London, Paris or Florence without wheeled vehicles of any kind. A great surprise in the Second World War was the discovery that the output of food per acre was four times greater in gardens and on allotments than it was on farms. For many people growing food on a small personal scale is a fulfilling part-time occupation.

More than 50 per cent of the world's inhabitants now live in cities, and in the wealthier nations over 90 per cent are urban dwellers. The trend is for more and more to leave the country for a new life in the city. I suspect that a well-run city uses less food and energy than a civilization of villages and isolated farms; certainly less than the distributed exurban communities that surround most of the developed world's towns today. We could almost assume that in the first world we are all the denizens of cities, and as such wholly dependent on regular supplies of food, water, raw materials and energy. Few in city houses or apartments ever go to the well for water or dig up potatoes to eat. Almost like a superorganism, the city grows and lays down, like the roots of a tree, its network of water, gas and sewer pipes and its electricity and communication cables. We only notice their existence when something goes wrong – a bad smell from the drains, cloudy water from the tap, or intermittent electric light and heat. To keep these supplies constant requires a powerful system of regulation, which almost always uses electricity. The sewers once carried their effluent using gravity alone to power the flow, but as the city grew pumps became necessary. These pumps and pumps for drinking water, with their control valves to regulate pressure, are all operated electrically from control stations. The underground and overground trains are electrically powered, as are the fuel pumps that fill the cars and lorries. Apartments and office buildings all have lifts and lighting dependent on the constant availability of electricity, and so do the mobile and landline phones that enable our ceaseless chatter. As well as all this there are the television and radio systems; and we play games, write letters and perform so many of the practical functions of modern life using personal computers – whoever heard of a gas or steam PC?

Such is the city dweller's total dependence on a constant, unvarying supply of electricity. Without it a city dies, and quickly – as you would without oxygen. The individuals in a city deprived of electricity live on a little while in frantic but aimless activity, just as do the cells of a newly dead body; within a week all that was alive is dead. The corpses are slowly repossessed by the natural world.

Have you ever thought what would happen to London, or any other major city, if there were no electricity for a week? This is what could happen if we put faith in green energy to run our lives. Imagine that wind turbines entirely replaced our ageing nuclear power stations, that we had taken the green message to heart and closed the last coal-fired power station to 'save the planet', and run our cities on wind and Russian gas. Sustaining a city requires constant and reliable supplies of energy. Enduring only intermittent supplies is not an option. This was well illustrated a few years ago on UK television by a BBC *Horizon* programme about the consequences of the failure of London's electricity supply. It showed as a drama what would happen to the city and its people if electricity were cut off for one week. The imagined cause was a disaster in Europe that disabled the supply of gas through pipelines beneath the North Sea, coinciding with a cold anticyclonic spell of weather that made wind power useless. I knew before the programme that urban life was greatly dependent upon electricity, but I did not know how many of our daily activities could not proceed without it. The programme revealed the vulnerability of sewerage and water supply systems, and how both depend on electricity for their operation; so do traffic lights, the pumps that supply fuels in petrol stations, and supermarkets, especially their refrigerators. Street and home lighting, lifts in high-rise buildings, hospitals and policing – indeed almost everything is dependent on electricity. The programme showed how in the course of a week London degenerated into the condition of a camp filled with starving refugees. This was fiction, but from discussion with colleagues in the energy and power transmission industries I have little doubt that it was near the truth.

Europe is peculiarly equivocal about nuclear energy. France is an example to all the world of a nation generating all its electricity from either hydro or nuclear energy. Much of the rest of Europe, with the notable exceptions of Sweden, Finland and other eastern Baltic nations, have been persuaded unwisely to reject nuclear energy in favour of an unsustainable mix of gas, coal and inefficient alternative energy schemes. The denial of nuclear energy on mainland Europe may be less serious because they have apparently no aversion to burning coal, possess a modest supply of hydroelectricity, and have faith in a constant supply of Russian gas. We in the UK have been pressured to use inefficient and insecure energy sources through the legal power of the European Union; they are ruinously expensive and if pursued will lead to our downfall. It costs us dearly in taxes to support them and discourages investment in nuclear energy which, without the need for subsidy, produces electricity more reliably and at a third the cost of wind energy. We thought when our leaders signed up to the renewables obligation that we were being truly green, and paid willingly; we did not see at the time that we were being conned into supporting a de facto European Common Energy Policy (CEP), mainly for the benefit of the German and Danish renewables industries. The existing Common Agricultural Policy (CAP) was introduced early into the European Community by President de Gaulle – unashamedly mainly for the benefit of French farmers. The main donors were Germany and later the UK when it was finally admitted to what was in 1973 an exclusive club. The CAP has remained beneficial to France and to Southern European nations and Ireland; Germany and the UK have remained the donors. Understandably Germany, which had borne this burden, felt that some restitution was needed. The Europeans have behaved sensibly according to their beliefs, but I think that British politicians failed to see how harmful both the CAP and the CEP would be to our nation. Foolishly we signed an agreement that committed us to powering our nation from an expensive, inefficient and unproven source. Alternative green energy could never supply our electricity needs now, and certainly not when, as seems possible, our population grows to 100 million. All of us in Europe, but especially the UK and Ireland, should abandon this romantic but foolish dream of relying on alternative energy. I applaud our present Prime Minister, Gordon Brown, for having the strength and wisdom to start rebuilding nuclear energy. It must have taken guts to go against the political pressures from Europe and those members of his party still reliving the fun of marching to Aldermaston proclaiming the need to make Britain a nuclear-free zone. He will find it much more difficult to repeal the

Luddite legislation that could delay by ten years attempts to rebuild our nuclear industry.

If the Earth does move to or near the hot state, more than 4°C hotter than now, only a limited area of land will be available for the natural ecosystems to share with us. It would probably be unwise of us to take more than 30 per cent of this area for ourselves, and to allow for expansion and mistakes it would probably be better to aim at no more than 10 per cent. Failure to keep the natural ecosystems of the land would leave the Earth's self-regulation entirely to the ocean ecosystems, which are, in a hot world, partially disabled by the formation of a warm top layer deprived of nutrients. A hi-tech, compact civilization would have these advantages: food synthesis would lessen its impact on the planet, and the widespread desert of this calorific planet would be an ample provider of solar electricity. Such a civilization gives us the chance to cease being a burden on Gaian regulation, and time to learn how to complement it. A high standard of living with women empowered and well educated would perhaps provide an automatic curb to population growth. If this were global in extent, disruption by war might be less likely.

5

Geoengineering

There are signs that we can treat global heating by engineering or other means. We have proved that our unscheduled and unintended experiment of adding large quantities of carbon dioxide to the air by burning carbon fuel heated the planet, and we now know that it was a mistake. Does this mean that we can cure global heating by adding some other gas or material that does the opposite and cools? Scientists, including me, think that we may have little option but to try; but surely it is much better to try as a planned experiment than as a panic response to, for example, the simultaneous flooding of several coastal cities.

If geoengineering is defined as purposeful human activity that significantly alters the state of the Earth, we became geoengineers soon after our species started using fire for cooking, land clearance, and smelting bronze and iron. There was nothing 'unnatural' in this; other organisms have been massively changing the Earth since life began 3.5 billion years ago. Without oxygen from photosynthesizers, for example, there would be no fires.

Organisms change their world locally for purely selfish reasons: if the advantage conferred by the 'engineering' is sufficiently favourable it allows them, their progeny and their environment to expand until dominant on a planetary scale. Our use of fires as a biocide to clear land of natural forests and replace them with farmland was our second act of geoengineering. Third was industry for the last 200 years. Together these acts have led us and the Earth to evolve to its current state. As a consequence, most of us are now urban and our environment is an artefact of engineering. During this long engineering apprenticeship we changed the Earth, but until quite recently, like the photosynthesizers, we were unaware that we were doing it; still less were we aware of the adverse consequences.

It might seem that the fourth assessment report of the IPCC, by more than a thousand of the world's most able climate scientists, who have worked on it since 1991, would provide us with most of what we need to know to ameliorate adverse climate change. Unfortunately it does not, and many climate scientists would acknowledge that their conclusions so far are tentative. The gaps that exist in knowledge about the state of the ocean, about that part of the Earth's surface that is ice, the cryosphere and even the clouds and aerosols of the atmosphere make prediction unreal. The response of the biosphere to climate and compositional change is even less well understood. We may soon need geoengineering applied empirically, because careful observation and measurement show that even today some parts of climate change, for example sea-level rise, are happening faster than the gloomiest of the forecasts.

GEOENGINEERING TECHNIQUES

Geoengineering methods fall into three main categories: physical means of amelioration such as the manipulation of the planetary albedo (the amount of sunlight reflected back to space); biological geoengineering that includes tree planting, the fertilization of ocean algal ecosystems with iron, the direct synthesis of food from inorganic raw materials and the production of biofuels; and, finally, active or Gaian geoengineering that involves the use of the Earth's ecosystem to power the process, or to change the nature of climate feedback from positive to negative. In connection with this idea I will also briefly describe the proposal that oceans be fertilized to encourage algal growth by mixing into the surface waters nutrient-rich water from below the depth which separates the warm surface water from the colder water below.

The most talked-about procedure for changing the Earth's heat balance by reflecting solar radiation back to space is the introduction of an aerosol of sulphuric acid droplets into the stratosphere. This was first suggested by the Russian climatologist Budyko in the 1970s and the pros and cons of it have been discussed since by distinguished scientists including Robert Dickinson, Paul Crutzen, Robert Charlson, and Meinrat Andreae and Ken Caldeira. Sulphur dioxide oxidizes rapidly in the stratosphere and the end product is sulphuric acid in the form of tiny droplets. These droplets are small enough to float in the stratospheric air like smoke and take three years to settle out. The strongest argument in favour is the fact that the volcano Pinatubo injected 20 million tons of sulphur dioxide into the stratosphere. Atmospheric global warming appeared to halt for the next three years. Apart from the severe local damage sustained in the Philippines when the volcano erupted, there do not seem to have been environmental changes significant enough to rule out the use of sulphur compounds for geoengineering. Many environmental scientists oppose the idea on the grounds that it would encourage business as usual and the continued emissions of carbon dioxide. Moreover, while the air temperature might be cooler, the increased carbon dioxide in the air would continue to damage ocean eco-systems through ocean acidification. I agree with this analysis but think that amelioration of this kind should be regarded as equivalent to dialysis as a treatment for kidney failure. It is valuable as a way to buy time, to survive until something better is available. Who would refuse dialysis if death is the alternative? It might be thought that the millions of tons of sulphuric acid aerosol would add significantly to the acidity of the ocean. This is the least of our worries because the quantity of sulphuric acid from the aerosol is tiny compared with the acidification from carbon dioxide, which when dissolved in the sea exists as carbonic acid.

Proposals for the introduction of sulphur into the stratosphere have inspired some ingenious inventions. Release from commercial aircraft when they are flying at suitable stratospheric altitudes would be the easiest, and chartered flight-refuelling tankers could be modified for this use. If 200 tons of sulphur could be lifted to the stratosphere by such planes, 100,000 flights would be the equivalent of Pinatubo. The chemical compounds that could be the precursors of a sulphuric acid aerosol include sulphur dioxide, hydrogen sulphide, carbon disulphide or slurry of colloidal sulphur. The first three are dangerously toxic to carry on a passenger plane and irritating or unbearably smelly. One or two tons of sulphur compound released by the majority of commercial aircraft would produce a substantial aerosol in a few years. It may not matter much where the release occurs. Large volcanoes in places as distant as Iceland, Indonesia and North America have all been followed by notable worldwide cooling.

Paul Crutzen and Ken Caldeira have both proposed other methods for introducing sulphur to the stratosphere. Crutzen proposed the use of military hardware: guns that fired shells laden with sulphur compound. Caldeira suggested a lightweight plastic pipe carried up by a balloon as a conduit for (probably) hydrogen sulphide since it is the lightest of sulphur carriers. If this treatment is prescribed we need some effective geo-clinical trials before it can be deployed. It is argued that the sulphate aerosol could worsen ozone depletion by chlorine and bromine compounds still in the air. I doubt if this argument, even if true, should be used to stop sulphate aerosol cooling. Ozone depletion, once a serious global problem, now pales in severity before global heating and if it occurs should be regarded as a side effect of an otherwise useful treatment.

Lowell Wood and others have proposed the use of a sunshade in orbit around the sun and in synchrony with the motion of the Earth. There is a natural point of stability that would allow the sunshade to stay in position with a minimum expenditure of energy. The shade would be a diaphragm made of fine carbon-fibre mesh that would be spun into a disc ten or more miles in diameter. It would disperse a few per cent of the sunlight incident upon the Earth. In principle it could work, but so far there seems to be little interest by agencies with the large funds that would be needed. It is in the same category as proposals to use space technology to prevent the Earth encountering hits from space debris of natural origin, and would have to be run by a large national or international agency.

Another way to increase the Earth's reflectivity is to make low-lying clouds over the oceans. These would be the artificial equivalent of natural marine stratus clouds. John Latham of the NCAR has joined with Steven Salter and pioneered the design of simple devices that make large numbers of cloud condensation nuclei (CCN) by spraying sea water. Because this approach has far fewer potential adverse side effects than the stratospheric aerosols, it should be tried on a sufficient scale to assess its worth.

It seems there is no shortage of geoengineering methods to offset global heating. Used alone they are no cure, since carbon dioxide would continue to increase and do damage in other ways than heating, but they could usefully provide a stay of execution while a more permanent treatment is developed.

SEQUESTERING CARBON DIOXIDE

The next class of geoengineering schemes are based on ways to remove carbon dioxide either from effluents of power stations and other large emitters, or even directly from the air. A great deal of research is now under way, mostly by energy companies, to produce an economic procedure for scrubbing carbon dioxide from furnace effluents, and so far it looks promising but would probably double the cost of electricity produced in this clean way; engineering development and economies of scale have a reasonable chance of reducing this penalty. Having sequestered the carbon dioxide it is far from easy to dispose of. How can we bury all of the carbon dioxide emitted by the large energy companies? Emissions total about 30 gigatons a year and the large producers emit less than a third of this, say 10 gigatons; if 10 gigatons were sequestered it would at best only slow global heating in the long term. To usefully reduce the stress of global heating most of the excess carbon dioxide would have to be removed from the carbon cycle and this can never be done by collection from large industrial sources only. The other difficult problem is the disposal of gigatons of carbon dioxide gathered this way. Underground burial in used oil and gas reservoirs is possible and is being pursued in Norway. We will have to wait and see how well and how economically it can be done. The bulk burial of carbon dioxide in underground reservoirs does have a special hazard. Because the gas is denser than air any large escape would pool at the surface and asphyxiate anyone in the pool. An event of this kind occurred naturally in Africa when carbon dioxide under pressure below a volcanic lake spilled over into villages along a valley: the death toll was large.

The most exciting idea among those for taking away carbon dioxide is Klaus Lackner's proposal of artificial trees. In essence it involves the use of readily available rock or soil to react directly either chemically or biochemically with atmospheric carbon dioxide and have as a product an easily disposable or, even better, a usable material. One example would be ground serpentine rock, a fairly common igneous rock that can consist of as much as 50 per cent magnesium oxide. The product, magnesium carbonate, is a stable white powder and could be used in building material or as a component of cement. A clear description of Lackner's ideas is in Wally Broecker's *Fixing Climate*, written with Robert Kunzig.

Tree planting would seem to be a sensible way to remove carbon dioxide naturally from the air, at least for the time it takes for the tree to reach maturity. But in practice the clearance of forests for farmland and biofuels is now proceeding so rapidly that there is little chance that tree planting could keep pace. Forest clearance has direct climate consequences through water cycling and atmospheric albedo change and is also responsible for much of the carbon dioxide emissions. Agriculture in total has climatic effects comparable with those caused by fossil fuel combustion. For this reason it would seem better to pay the inhabitants of forested regions to preserve their trees than plant new trees on cleared ground. The charity Cool Earth exists to gather funds for this objective as does the Prince's Forest Trust in the UK. It is insufficiently appreciated that an ecosystem is an evolved entity comprising a huge range of species from microorganisms, nematodes and invertebrates to small and large plants and animals. While natural ecosystems have the capacity to evolve with climate change, plantations can easily die.

Oceans cover over 70 per cent of the Earth's surface and are uninhabited. In addition, most of the ocean surface waters carry only a sparse population of photosynthetic organisms, mainly because the mineral and other nutrients in the colder water below the thermocline (the lower boundary of the warm surface waters) do not readily mix with the surface layer. Some essential nutrients such as iron are present in suboptimal abundance even where other nutrients are present. This led to the suggestion by John Martin that fertilization with the trace nutrient iron would allow algal blooms to develop that would cool the Earth by removing carbon dioxide . The most recent research suggests that this approach may hold promise despite early disappointments.

In 2007 Chris Rapley and I suggested the use of a system of large pipes held vertically in the ocean surface to draw up cooler, nutrient-rich water from just below the thermocline. The intention was to cool the surface directly, to encourage algal blooms that would serve to pump down carbon dioxide and also to emit gases such as dimethyl sulphide, volatile amines and isoprene that encourage cloud and aerosol formation. The pipes we envisaged would be about 100 metres in length and 10 metres in diameter, held vertically in the surface waters and equipped with a one-way valve. Surface waves of an average height of one metre would mix in five tons of cooler water per second.

Our intention was to stimulate interest and discussion on physiological techniques that would use the Earth system's energy and nutrient resources to reverse global heating. We do not know if the proposed scheme would help restore the climate but we have discovered that such pipes are already in use commercially to improve the quality of ocean pastures for fishing. The reaction from the scientific community was an almost instant rejection on the grounds that their use would release carbon dioxide from the lower waters to the atmosphere. We were aware of this drawback but thought it reasonable to expect that algal growth following the mixing might take down more carbon dioxide than was released. The next step would be the experimental deployment of the pipes, observations and measurements.

If any of these ocean fertilization schemes work then their value could be enhanced by harvesting the algae, extracting food and fuel and then burying the waste in the deep ocean as heavier-than-water pellets. This would remove a sizeable proportion of the carbon photosynthesized and place it as an insoluble residue on the ocean floor. The temperature of the deep ocean is close to 4°C and the residence time of water there at least a thousand years. The buried carbon would effectively be out of circulation. It might be possible also to bury land-based agricultural waste at these deep ocean sites.

THE PROMISE OF THE BURIAL OF ELEMENTAL CARBON

By far the most promising and practical way to take the excess carbon dioxide from the air is to ask Gaia to do it for us. All of the schemes for sequestering carbon dioxide require us to use energy (and most probably this would be fossil-fuel energy) to do it. Even the sum total of all our pollution output of carbon dioxide is still small compared with the turnover from the Earth. We emit 30 giga-tons a year but Gaia emits 550 gigatons; so if Gaia can balance this huge quantity can we not persuade her to do better? I think that we can, by making a small change to the carbon cycle. Ordinarily 99.9 per cent of the carbon that the photosynthesizers take from the air is returned by consumers that oxidize it back to carbon dioxide or convert it to methane. The earliest reference I could find to burying elemental carbon as a remedy for global heating was by Johannes Lehmann in a commentary article in *Nature* in 2007. The idea of converting agricultural waste into 'char' (char is closely similar to charcoal but needs a separate word to distinguish it be-cause it is not a fuel) is now the subject of research and development. The conversion of agricultural waste into char at one stroke changes the natural release of 99.9 per cent of the carbon of the waste as carbon dioxide and methane into a release of only 10 to 30 per cent, a vast improvement on its direct use as a source of biofuel.

If the bulk of agricultural waste were turned to char on farms it could be buried in the soil and that way the crop plants photo-synthesizing solar energy would have taken the carbon dioxide from the air for us. It is much more economic to use the huge and free power of photosynthesis to remove carbon dioxide than to use manufactured energy. It might even be possible to convert waste from algal farms in the ocean into char and let it fall to the sea floor. We would be denying the natural consumers of algae their food but in the long run they would benefit because if global heating is allowed to proceed as now there will be few producers or consumers left in the oceans.

It is not commonly known that char is almost completely inert. Neither atmospheric oxidation nor the action of micro-organisms returns it to the air as carbon dioxide. This makes it safe to bury in the soil or in the ocean. So far it is the only realistic proposal by which we have even a chance of restoring the Earth to the state it was in before we started using fossil fuel. It even has a bonus in that the act of making charcoal provides a benign form of biofuel as a by-product. Pooran Desai and Sir Ghillean Prance first drew my attention to this promising idea and I am indebted to David Wayne who let me see the unpublished text of his article on 'The Biochar Opportunity'. Research into its practical and engineering development is now under way at Shell Research Ltd.

Another amelioration technique is the direct synthesis of food from carbon dioxide, nitrogen and trace minerals. Now that food is abundant it seems an otiose proposal but it would release land that could return to its former natural state with the capacity to regulate the climate.

Although not usually thought of as geoengineering, the synthesis of food and liquid fuels from carbon dioxide and water, using high temperature nuclear reactors to produce the carbon compound feedstock, is an effective way to remove carbon dioxide from the air.

GEOPHYSIOLOGY

In Chapters 2 and 6 I compare the Earth system to the physiology of an animal and describe how normally it stays in homeostasis and how the Earth system is dynamically stable but has strong feedbacks; because of the huge turnover of the Earth's surface by living organisms, its response to change is like that of a living organism. Yet even wholly physical models of the Earth system are non-linear, often because the properties of water set critical points during warming and cooling. For example, the phase change from ice to water is accompanied by an albedo change from 0.8 to 0.2, and this strongly affects climate, as Budyko first described. This feedback is now influencing climate change and will continue until the ice melts. There are other purely physical feedbacks in the system: the ocean surface stratifies at 12–14°C; the rate of water evaporation from land surfaces becomes a problem for plants at temperatures above 22–25°C; and atmospheric relative humidity has a large direct effect on the size and effective albedo of aerosols. The combined effect of feedbacks involving the physical and biological responses of the Earth can be the source of large discontinuities in climate and chemical composition. The existence of these discontinuities is often accompanied by marked hysteresis – that is, a reluctance to move from one state to the other even when pushed beyond a tipping point.

I described in Chapter 2 a model of a planet with a land surface occupied by plants and the ocean a habitat for algae – a model which showed strong self-regulation of its temperature. But with rising carbon dioxide or heat flux there was a sudden 5°C rise that took place at 450 ppm of carbon dioxide; there was marked hysteresis and reducing temperature or heat flux did not immediately restore the state that had existed before the discontinuity. The behaviour of this simple geophysiological model and the Earth's recent climate history revealed by ice-core analysis indicates a climate and atmospheric composition that fluctuates suddenly, as would be expected of a dynamic system with positive feedback. An engineer or physiologist looking at the historic response of the Earth system would think it unwise to assume that climate change can simply be reversed by reducing emissions or by geoengineering.

The long-term history of the Earth suggests the existence of hot and cold stable states that geologists refer to as the greenhouses and the ice houses. In between are metastable periods like the present interglacial. The best known hothouse happened 55 million years ago, near the beginning of the period known by geologists as the Eocene. It was so called because it marked the dawn ('eos') of large mammals. The Eocene was already warm by present standards, and a geological accident caused the release of between 1 and 2 teratons of carbon dioxide into the air (a teraton is one million million tons). Putting this much carbon dioxide in the air caused the temperature of the temperate and Arctic regions to rise by 8°C, and of the tropics by between 5° and 8°C; and it took about 200,000 years for conditions to return to their previous states. Soon we will have injected a comparable quantity of carbon dioxide into the atmosphere, and the Earth itself may release as much again.

There is good evidence that temperature and carbon dioxide rose sharply in the Eocene event, but the cause remains uncertain. The two favoured speculations are the sudden release of a large volume of methane from its unstable entrapment in crystals called 'clathrates': methane is itself a powerful greenhouse gas but soon oxidizes to carbon dioxide. The other speculation concerns the incursion of molten lava beneath a petroleum deposit in the Arctic Ocean. The accident that caused the large rise of atmospheric carbon dioxide 55 million years ago is thought to have occurred more slowly than now: the injection of gaseous carbon compounds into the atmosphere may have taken place over a period of about 10,000 years, instead of about 200 years. The great rapidity with which we add carbon gases to the air may be as damaging as the quantity of them. The rapidity of the pollution gives the Earth system little time to adjust, and this is particularly important for the ocean ecosystems: the rapid accumulation of carbon dioxide in the surface water is making them too acid for shell-forming organisms. This did not appear to happen during the Eocene event, perhaps because there was time for the more alkaline deep waters to mix in and neutralize the surface ocean. Despite the large difference in injection times of carbon dioxide, the change in temperature of about 5°C globally may have occurred as rapidly 55 million years ago as it may soon do now. The time it takes to move between the two system states is likely to be set by the properties of the system more than by the rate of addition of radiant heat or carbon dioxide.

There are differences between the Earth 55 million years ago and now. The sun was 0.5 per cent cooler and there was no agriculture anywhere, so that natural vegetation was free to regulate the climate. Another difference was that the world was not then experiencing global dimming – the 2 to 3 degrees of global cooling caused by the atmospheric aerosol of man-made pollution.

PLANETARY MEDICINE AND ETHICS

What are the planetary health risks of geoengineering intervention? Nothing we do is likely to sterilize the Earth, but the consequences of planetary-scale intervention could hugely affect humans. Putative geoengineers are in a similar position to that of physicians before the 1940s. In his book *The Youngest Profession* the physician Lewis Thomas beautifully described the practice of medicine before the Second World War. There were only five effective medicines available: morphine for pain, quinine for malaria, insulin for diabetes, digitalis for heart disease and aspirin for inflammation, and very little was known of their mode of action. For almost all other ailments there was nothing available but nostrums and comforting words. At that time, despite a well-founded science of physiology, we were still ignorant about the human body or the host–parasite relationship it had with other organisms. Wise physicians knew that letting nature take its course without intervention would often allow natural self-regulation to make the cure. They were not averse to claiming credit for their skill when this happened. I think the same may be true about planetary medicine: our ignorance of the Earth system is overwhelming and intensified by the tendency to favour model simulations over experiments, observation and measurement.

Global heating would not have happened but for the rapid expansion in numbers and wealth of humanity; if we fail to curb global heating, the planet could massively and cruelly cull us, in the same merciless way that we have eliminated so many species by changing their environment into one where survival is difficult. But before we start geoengineering we have to ask: Are we sufficiently talented to take on what might become the onerous permanent task of keeping the Earth in homeostasis? Consider what might happen if we start by using a stratospheric aerosol to ameliorate global heating – even if it succeeded it would not be long before we faced the additional problem of ocean acidification. This would need another medicine, and so on. We could find ourselves enslaved in a Kafkaesque world from which there was no escape. The alternative is the acceptance of a massive natural cull of humanity and a return to an Earth that freely regulates itself.

Whatever we do as geoengineers is unlikely to stop dangerous climate change or prevent death on a scale that makes all previous wars, famines and disasters small; but to continue 'business as usual' could be worse and would probably kill most of us during the century. We have to consider seriously that, as with nineteenth-century medicine, the best option may be kind words and painkillers, but otherwise doing nothing and letting Nature take its course.

The usual response to such bitter realism is defeatist: 'Then there is no hope for us, and we can do nothing to avoid our plight?' This is far from true. We can adapt to climate change, and this will allow us to make the best use of the refuge areas of the world that escape the worst heat and drought. We have to marshal our resources soon, and if a safe form of geoengineering can buy us a little time then we must use it. Parts of the world, such as oceanic islands, the Arctic basin and oases on the continents, will still be habitable in a hot world. We need to inhabit them and see that they have sufficient sources of food and energy to sustain us as a species.

During the early Eocene global heating there was no great extinction of species and this may have been because life had time to migrate to the cooler regions near the Arctic and Antarctic and remain there until the planet cooled again. This may happen again and humans, animals and plants are already migrating. Scandinavia and the oceanic parts of Northern Europe such as the British Isles may be spared the worst heat and drought that global heating brings. This puts a special responsibility upon us to survive but also where possible give refuge to climate refugees from more distant places.

Perhaps the greatest value of the Gaia concept lies in its metaphor of a living Earth, which reminds us that we are part of it and that our contract with Gaia is not about human rights alone, but includes human obligations.

6

The History of Gaia Theory

The idea of an Earth system science, a self-regulating Earth with the community of living organisms in control, came into my mind at the Jet Propulsion Laboratory in California in September 1965. The first paper to mention it was published in the *Proceedings of the American Astronautical Society* in 1968. The title of the paper was 'Planetary Atmospheres: Compositional and other changes associated with the presence of Life'. It was an almost unnoticed paper, and was mainly concerned with atmospheric analysis as an extraterrestrial life-detection experiment. But here are two paragraphs from the paper that illustrate how the Gaia hypothesis arose in the period before it received its name:

If the atmosphere of the Earth is a biological contrivance, then it is reasonable to consider that the components are maintained at an optimum or near optimum composition, for the ecosystem. For example, the Earth's climate is strongly dependent on the atmospheric pressure, that is, the total amount of oxygen and nitrogen, and on the concentration of infra-red absorbency gases such as carbon dioxide and water vapour. The concentration of these components is directly or indirectly under biological control. It may not therefore be an unreasonable speculation to consider the possibility that the Earth's climate is also maintained near an optimum for the ecosystem.

It is interesting to ask, why is the oxygen concentration maintained at 21 per cent? It is a fact that the energy required for the ignition of organic compounds changes by about 70 per cent for each 1 per cent change in oxygen concentration at the atmospheric level. Life at even 25 per cent oxygen might be very uncomfortable, especially for trees. The removal of oxygen by grass or forest fires may set the upper limit of 21 per cent but it seems more likely that oxygen is actively controlled at a safe maximum.

So the Gaia concept was born at the peak of the New Age – contemporary with Woodstock and the Beatles, which perhaps accounts for why so many scientists still regard it as part of the plethora of New Age nonsense which was around at the time. But not all of us were hippies with our rock chicks. There was the space programme that culminated with the moon landings, a surge of planetary exploration by orbiting satellites, and the discovery of DNA and the genetic code. The 1960s saw the near catastrophic confrontation between the superpowers over missiles sited on Cuba, and the end of segregation in the USA and much violent political change; it was a time of painful conflict between old and new views of the world.

Apart from the coincidence of its birth with the New Age, Gaia science was far too revolutionary an idea to be immediately accepted, and I should not have expected this until a substantial quantity of evidence and theory had been gathered; it was not in fact until thirty-six years later, in 2001, that the concept received partial public recognition. I was tempted in the 1990s to admit defeat and settle for some anodyne phrase such as 'Earth system science' or a portmanteau science such as biogeochemistry. But, as Fred Pearce in an article in the *New Scientist* in 1994 so well describes, I listened to my friends, Jonathon Porritt, Mae Wan Ho and Mary Midgley. It might have pacified the 'scientifically correct' for me to have dropped Gaia but it would have been surrender to a conformity that I knew was wrong. I am glad that I have stayed faithful to the name Gaia for more than forty-three years. Perhaps if I had never met Bill Golding and had let my ideas be called the dull and uninspiring 'Earth system hypothesis', the name hinted at in my 1968 paper, biologists would never have read the subsequent papers that so irritated them. Science might then have known as much as thirty years sooner the true nature of the climate threat we now face and had time to take appropriate action.

In science an idea is promoted to the level of a hypothesis when some aspect of new, reliable evidence needs an explanation. For Gaia theory the new evidence was the detailed analysis of the composition of Mars' and Venus' atmospheres revealed from the infra-red spectra of the planets gathered at the Pic de Midi observatory in France by the husband and wife team of astronomers Pierre and Janine Connes. The Connes released this information in September 1965 and we received it at JPL. Before this I had argued that the easiest way to see if Mars had life was simply to measure the chemical composition of its atmosphere. My argument was that if the planet had no life then the atmosphere would be close to chemical equilibrium; that is to say, no energy would come from the gases of the atmosphere reacting together. By contrast if the planet bore life then the organisms would be obliged to use the atmosphere, the only mobile medium on Mars, as a source of raw materials and as a place for waste disposal. Such a use of the atmosphere would make it recognizably different from the equilibrium atmosphere of a dead planet. The spectroscopic data gathered by the Connes about Mars and Venus showed that their atmospheres were almost entirely composed of carbon dioxide, and that oxygen, nitrogen, etc. were at low levels. These planetary atmospheres lacked any available chemical reactivity and were therefore near to chemical equilibrium, and according to my hypothesis there was no abundant life on them.

The Earth was the control planet, the one we were sure held life, and its atmosphere is profoundly at disequilibrium. We have oxygen and methane simultaneously present at 21 per cent by volume and 1.5 parts per million respectively. In the presence of sunlight methane oxidizes, and after only about ten years 67 per cent of it has gone. Yet methane has been fairly constant, as ice-core analyses prove, for the past million years, as has oxygen. Such constancy implies a degree of disequilibrium with an astronomical improbability. That is, for such constancy to happen by chance is infinitely improbable. Similar improbabilities apply to the presence of the other gases – nitrogen, carbon dioxide, nitrous oxide, and so on. The only exceptions are the rare gases such as argon, helium and xenon that are chemically un-reactive. Since all the gases other than the rare gases are either made by organisms or processed by them, I could offer the Gaia hypothesis which stated that the Earth's atmospheric composition is kept at a dynamically steady state by the presence of life; moreover if organisms could affect atmospheric composition then maybe they could regulate the climate of the Earth to keep it favourable for life. It was known in the 1960s that the sun had warmed by at least 25 per cent since life began 3.5 billion years ago, and regulation would have been needed to retain habitability. The hypothesis was published in peer-reviewed papers in the late 1960s and early 1970s.

In the early 1970s I was invited to visit Lynn Margulis at her Boston laboratory. Lynn was no stranger to controversy and fought hard to establish the endosymbiont hypothesis – now happily conventional wisdom, but which in its time seemed as embattled as was the Gaia hypothesis. In certain ways we formed a small revolutionary cell in a world of conservative Earth and life scientists. Lynn made a huge contribution to the concept of Gaia by stressing the importance of the micro-organisms in the evolution of our planet. She made it clear that for about 2 or 3 billion years from its start, the biota – that is, all forms of life on Earth – were micro-organisms. Only in the last 500 to 900 million years have multicellular organisms played a part. Without Lynn I would never have met those formidable Earth scientists who opposed what was then the Gaia hypothesis, H. D. Holland, professor of Geology at Harvard University, and James Walker, then at Yale University. Both of them rejected Gaia, but as good scientists they were prepared to argue about it. As so often with battles, after the war is over the combatants enjoy sharing experiences, and by doing this Dick Holland has become a friend who agrees to differ, generously enriching the meetings that Sandy and I have held at Green College in Oxford. Holland's main criticism, and one expressed in his splendid book *The Chemical Evolution of the Atmosphere and Oceans*, is simply that Gaia is not needed to explain the geochemistry of the Earth – Earth science alone is enough.

James Walker, Jim Kasting and P. B. Hayes were the first to propose a mechanism for stabilizing the Earth's temperature and carbon dioxide abundance in the early 1980s, but like Holland they held that regulation could be explained by geochemistry alone. They used the well-established fact that there is only one source of carbon dioxide – volcanoes and tectonic processes; and only one sink – the removal of carbon dioxide from the air by its reaction when dissolved in rainwater with rock that contains calcium silicate (basalts and granites). The products of this reaction are the water-soluble compounds calcium bicarbonate and silicic acid, and these travel in the groundwater and rivers to the oceans. They explained regulation by noting that when the temperature is high more water will evaporate from the ocean and more rain will fall, which will increase the rate of the rock weathering reaction and so lower the abundance of carbon dioxide in the air. This is a dynamic process with inherent negative feedback that could stabilize both carbon dioxide and temperature. They were therefore proposing that a lifeless Earth could regulate its temperature at levels habitable for organisms.

When I first heard Walker describe this mechanism at a Dahlem Conference in Berlin in the early 1980s it certainly sounded plausible, and afterwards in conversation he said that his motivation was mainly to show that Gaia was not needed for self-regulation and that geochemistry could do it alone. At the time Andrew Watson, now a professor of biogeochemistry at the University of East Anglia, was working with me as a postgraduate student and it occurred to us that in real life rocks always have lichen and other organisms on their surfaces and, more importantly, the soil where fragments of the rocks exist is a rich ecosystem in its own right and has an internal atmosphere up to thirty times richer in carbon dioxide than the air. The rate of weathering in these circumstances could be much greater than on exposed bare rock. Tyler Volk and D. W. Schwartzman published a paper in *Nature* in 1989 that confirmed our speculation by a direct in vitro experiment. The organisms of an ecosystem respond to a rise of temperature by growing faster; the plants draw down carbon dioxide from the air; and in the soil more carbon dioxide is produced by consumers. The flow of carbon dioxide from the air to the rocks is enhanced and weathering proceeds more rapidly, consequently the carbon dioxide produced, instead of adding to the atmosphere, is removed by weathering. Removing carbon dioxide from the air lowers temperature and the system settles at a dynamic equilibrium close to the optimum for plant growth. Organisms are also crucial in the oceans to return carbon dioxide brought down in the rivers as calcium bicarbonate into calcium carbonate which settles as sediments on the ocean floor. The whole process, one that might be called biogeochemical rock weathering, is a Gaian mechanism and seems much more likely to be the basis for real-world temperature regulation. But we owe a debt to Walker and his colleagues for pointing us in the right direction.

Proof that the Earth self-regulates carbon dioxide abundance and temperature had to wait until 2008, when the American scientists Richard Zeebe and Ken Caldeira published a paper in *Nature Geosciences* showing that the long-term record of the Earth's temperature and carbon dioxide abundance, deduced from measurements of gases in Antarctic ice cores, revealed self-regulation of both carbon dioxide and temperature for hundreds of thousands of years. This evidence, if confirmed, provides splendid support for Gaia theory, but the authors referred only to the purely geochemical Walker model as the mechanism for regulation.

There is no animosity involved in the arguments I have with American Earth scientists over whether the Earth's regulation is a matter of Gaia theory or geochemistry. In the fine book, *The Earth System*, written by Lee Kump, James Kasting and Robert Crane, the authors reveal the friendliness of our relationship. The reason for disagreement lies in the often reductionist and disciplinary nature of the Earth and life sciences. This makes it difficult to share ideas about Gaia. As I see it, to understand Gaia requires an instinctive familiarity with the dynamics of systems in action, and this is not a normal part of Earth or life science.

Geology can be a joyful profession, especially if you enjoy exploring and spending time in the wilderness. Some of the most enthralling journeys into the countryside I have made were in the company of the American geologist Robert Garrels. With his hammer he would chip a small fragment of rock from a cliff face and reveal its provenance, and then tell me how a few hundred million years ago there was a world of blazing heat and drifting sand where we stood, or of tundra at the edge of a vast glacier. For Earth scientists their world was satisfying until Gaia loomed to falsify or complicate their elegant explanations. The same is true of field biologists; no wonder Gaia is unpopular. Were it not for the deadly serious consequences of using the wrong theory, the disagreement would be no more than the normal slow progress of scientific understanding.

It is normal to debate a new hypothesis, so what went wrong? Why was the Gaia hypothesis thrown into the rubbish bin? The trouble started in 1979 when the Canadian biologist Ford Doolittle wrote his lively and well-written critique of Gaia. Interestingly, he chose to publish it in the American New Age magazine *Coevolution Quarterly*, edited by Stewart Brand. Scientists may pretend to deplore the New Age, but that does not stop them reading its publications and in no time Gaia's face was turned to the wall, especially in the neo-Darwinist community of scientists. Neither Lynn Margulis nor I could make a convincing defence – partly because, as we had stated it, the Gaia hypothesis was wrong. We had said that organisms, or the biosphere, regulated the Earth's climate and composition. Somewhat later, in his book *The Extended Phenotype*, Richard Dawkins showed that this was impossible. He said it so well and clearly that the subject was then regarded by the scientific community as closed. Richard Dawkins is an extraordinarily talented author and persuader and in his book he vented his scorn on the Gaia hypothesis with the powerful erudition that he now uses to censure theology. From then on it became impossible to publish any paper on it in a mainstream journal; the peer reviewers were convinced by Dawkins and other eminent biologists that Gaia was mere New Age fantasy. I was shocked by the rejections because previously I had found peer reviewers helpful and had rarely had a paper rejected by a journal. It seemed in the 1980s almost as bad as censorship until the editor of *Nature*, John Maddox, learnt that during his absence the paper that Andrew Watson and I had written on the Daisyworld model had been rejected. He wrote asking me to send the next paper on a Gaian topic to him personally and in confidence. He promised that if it were of the quality of the Daisyworld paper it would be published in *Nature*. He was true to his word and the next paper on the topic was one I wrote with Robert Charlson, Meinrat Andreae and Steven Warren on the connection between clouds, condensation nuclei, dimethyl sulphide and its source, ocean algae.

I accepted Dawkins' criticism that there was no way for life or the biosphere to regulate anything beyond the phenotype of its component individual organisms. So what on Earth was doing the regulating? I had no doubt that climate and chemistry were regulated, so what did it if not life? As I have explained earlier, the traditional

Earth scientists, led by James Walker and H. D. Holland, were sure that regulation was done by geochemistry and geophysics alone and that life was a mere passenger or at most a contributor. But the solid evidence of massive disequilibrium from atmospheric composition made their simplistic explanation impossible. The thermodynamics and kinetics of gas reactions make the simultaneous presence of oxygen and methane at their observed abundance, the existence of nitrous oxide and the low concentration of carbon dioxide wholly inexplicable by inorganic processes alone.

I was as near certain as a scientist can be that the argument for the existence of self-regulation drawn from atmospheric disequilibrium was correct; moreover by now evidence was available from the Earth that confirmed several of Gaia theory's predictions. To me it was obvious that Richard Dawkins' pure biology and the geochemists' pure chemistry were unable to explain the Earth. And then I wondered, what if the whole system of life and its environment tightly coupled did the job? In 1979 it occurred to me that the biologists' objections would collapse if the regulator could be shown to be the whole Earth system, made up from all of life, including the air, the oceans and the surface rocks, not just organisms alone. To prove this would require an experiment on the whole Earth. In fact this was happening through our own emissions of carbon dioxide: we were perturbing the system and eventually there would be evidence to prove whether or not it self-regulated according to the Gaia hypothesis. But, as mentioned earlier, it was not until 2008 that Richard Zeebe and Ken Caldeira used ice-core evidence to show that it was.

All that I could do in 1981 to test this idea was to compose the holistic model, Daisyworld. Just before Christmas that year I wrote the program of the model and ran it on a 9845 Hewlett-Packard desktop computer. In some ways this was the most important step in the history of Gaia theory. It expresses succinctly the mathematical basis of the theory and can and has been tested to see if it can be falsified.

I wrote a program that described in mathematical terms a self-regulating system made up from the climate of a simple flat planet, illuminated by a star like the sun, and on which there was a simple ecosystem of two daisy species evolving in a Darwinian fashion. This Daisyworld had a surface temperature determined by the proportion of radiant heat from its star that was absorbed or reflected into space and by the quantity of heat radiated away in the infrared. There were no greenhouse gases to complicate the climate and the surface reflection of sunlight was proportional to the area covered by dark or light coloured daisies, or by bare ground. The daisies did not grow below 5°C or above 40°C, and grew best at 22.5°C. The model was run by slowly increasing the heat output of the star in a manner similar to the increase from the sun that has happened since the Earth formed 4.5 billion years ago. As soon as some part of the planet reached 5°C, dark daisies began to grow because being dark they absorbed more heat. Soon daisy growth and surface temperature increased rapidly as the daisies spread, until the planet became too hot for further dark daisy growth. Now white daisies began to compete for space, and as the star further increased its heat output, light-coloured daisies took more space until they dominated the planetary surface. Finally the star's heat was too much for the light daisies; they died off and the planet rapidly increased its surface temperature and became uninhabitable. A characteristic of this kind of model is that it shows what physicists term hysteresis; that is, if run backwards from the final hot state by reducing solar heat, white daisies do not appear until a considerably lower temperature is reached. The same is true as the lifeless cold state approaches: dark daisies persist at lower solar heat than that needed for their first appearance.

When the model Daisyworld was run I was delighted to find that the whole system of life and its environment regulated temperature at a level close to the optimum for plant growth. For a model laden with non-linear differential equations it was astonishingly stable and well behaved. It kept the temperature close to ideal for daisies over a considerable range of solar radiation inputs, but when the star illuminating Daisyworld was too bright or too dim all life vanished: the model planet was alive at tolerable heat inputs but dead if the star was too hot or too cool. It is important to recognize that Daisyworld is the model of an emergent system on which climate and organisms are tightly coupled and evolve together.

Daisyworld is much more than a population biology model about the spread of daisy types on a planet; it is also a climate model. What made it special was that for the first time the growth and selection of plants was tightly linked in a dynamic model to their ability to affect the climate and be affected by it. It showed how such a system could keep surface temperature close to the optimum for plant growth over a wide rage of radiant forcing. Variations on the theme of Daisyworld were composed and are described in my book *The Ages of Gaia*. Andrew Watson was at the time researching Gaia theory with me; he is a more competent mathematician than I am and his insight into the subtleties of the model enriched our joint paper on it, published in the Swedish journal *Tellus* in 1973.

Daisyworld was like a stick thrust into a wasp nest: the angry buzzing of biologists out to sting it to death was deafening. Papers were published claiming to falsify Daisyworld – naturally there were few peer reviewers who objected to such anti-Gaian publications. None succeeded in their goal, and Daisyworld remains unfalsified. In 2002 an editorial in *Nature* commented that no simple model had irritated as many scientists as Daisyworld. In any other branch of science than biology the failure to falsify the Daisyworld model should have made Gaia theory worthy of further investigation in the 1980s. Moreover the failure of the falsifications should have been a warning that neo-Darwinist theory was flawed.

The critics kept saying, 'What about cheats?' They thought Daisyworld would be bound to fail if cheats – freeloading daisies that merely grew and did not perform regulation – were included. It was easy to add a cheat, a neutral-coloured daisy species that did nothing for regulation, and reduce the growth rate of the others for the energy spent making pigment, but when I did this the model worked as well as before. The neutral-coloured daisies were only selected by the system when regulation was not needed; when it was hot, light-coloured, heat-reflecting daisies were favoured, when cold only dark-coloured, heat-absorbing ones were chosen. Daisyworld is Darwinian: biologists had failed as Darwin's disciples to realize that organisms do not evolve independently of their environment – in fact organisms are part of a larger whole that includes the physical and chemical environment that they and other organisms change.

For me the long and seemingly never-ending battle for the recognition of Gaia theory was frustrating and disappointing; but, for all that, rancour was evenly balanced with humour. The most distinguished Darwinist biologists, William Hamilton and John Maynard Smith, both became friends in the late 1990s even though Maynard Smith had earlier referred in public to Gaia as 'an evil religion'. He came to stay briefly at Coombe Mill in 1996 and over dinner told us that in the 1970s, when the Gaia hypothesis first appeared, Darwinist biologists were arguing fiercely with other biologists who believed that evolution took place by group selection, not the selection of individual organisms. At that time he saw Gaia as profoundly anti-Darwinian and far worse than group selection: the hypothesis that the planet evolved as if it were a living organism was to Darwinists at the time an absurd idea. During this visit we had great fun together over the neo-Darwinist argument about whether a brave person should jump into a river to save either a direct blood relative or eight cousins. Sandy and I, both of us poor swimmers, felt that trying to save eight of our cousins was carrying logic a step too far.

Daisyworld has turned out to be a fruitful source of other models of the Earth. Mathematicians, among them Peter Saunders, Inman Harvey and James Dyke, have even found its mathematical basis worthy of study. Professor Saunders and the physiologist Johan Koeslag mapped the mathematical basis of Daisyworld on to a model for human diabetes. Tim Lenton has published many Gaian papers based on Daisyworld and has organized a series of conferences on Daisyworld and its mathematical implications. These have been both popular and successful. Daisyworld itself evolved, in two different ways. First, it became a more comprehensive set of biological models, where instead of just two fixed daisy species there were up to one hundred different plant species and also herbivores and carnivores present at three trophic levels. This work is summarized in my paper 'A Numerical Model of Biodiversity' in *Philosophical Transactions of the Royal Society* in 1992. These models, including one where the organisms could mutate spontaneously, go far towards explaining the relationship between biodiversity and regulation. My friends Stephan Harding and Tim Lenton have carried them considerably further.

As a scientist closer to physics than other disciplines, I knew that the value of a theory is judged by the accuracy of its predictions and its capacity to resist falsification. By the early 1990s Gaia theory had made ten predictions and eight of them had been confirmed or at least become generally accepted. Furthermore, as physicists know, the predictions of good theories lead to outbursts of new scientific research. This has been especially true of the research stimulated by the prediction of a link between the biological production of dimethyl sulphide in the ocean, clouds in the atmosphere, the Earth's radiation balance and climate regulation. The paper on clouds, algae and climate by Charlson, Lovelock, Andrea and Warren was published in *Nature* in 1987, and its conclusions are usually referred to as the CLAW hypothesis. Since then hundreds if not thousands of papers have been published on researches it stimulated. Professor Liss of the University of East Anglia and I published a paper in 2007 in *Environmental Chemistry* summarizing the progress of the CLAW hypothesis, and concluded that the mechanism proposed was observable only in the unpolluted southern hemisphere. Sulphur pollution in the northern hemisphere is now as much as ten times greater than the natural output from algae, and obscures any effect the algae might have.

Prediction	Test	Result
Mars is lifeless (1968)	Atmospheric compositional evidence shows lack of disequilibrium	Strong confirmation, Viking mission 1975
That elements are transferred from ocean to land by biogenic gases (1971)	Search for oceanic sources of dimethyl sulphide and methyl iodide	Found 1973
Climate regulation through biologically enhanced rock weathering (1973)	Analysis of ice-core data linking temperature and CO_2 abundance	Confirmed 2008, by Zeebe and Caldeira
That Gaia is aged and is not far from the end of its lifespan (1982)	Calculation based on generally accepted solar evolution	Generally accepted

Prediction	Test	Result
Climate regulation through cloud albedo control linked to algal gas emissions (1987)	Many tests have been made but the excess of pollution interferes	Probable for southern hemisphere
Oxygen has not varied by more than 5 per cent from 21 per cent for the past 200 million years (1974)	Ice-core and sedimentary analysis	Confirmed for up to 1 million years ago.
Boreal and tropical forest are part of global climate regulation	Models and direct observation	Generally accepted
Biodiversity a necessary part of climate regulation (1992)	By models but not yet in the natural ecosystems	Jury still out
The current interglacial is an example of systems failure in a physiological sense (1994)	By models only	Undecided
The biological transfer of selenium from the ocean to the land as dimethyl selenide	Direct measurements	Confirmed 2000, Liss

Table 3. *The test applied to some of the predictions from Gaia, and the results.*

The next important step in the history of Gaia was the Amsterdam Declaration, made at a meeting of the European Geophysical Union in 2001, where more than a thousand scientists signed a statement that began, 'The Earth System behaves as a single, self-regulating system comprised of physical, chemical, biological and human components.' My friends said, 'At last Gaia is recognized as science'; but I knew that it still had some way to go, that the declaration was incomplete, and that Gaia theory would not truly be part of science until such a declaration included in addition a scientifically acceptable rendition of the idea that the goal of self-regulation is the maintenance of habitability. Earth and life scientists at Amsterdam had not realized how ambiguous it is to speak of self-regulation without specifying the aim, goal or set point of the system. Because science is still deeply in thrall to the rational Cartesian logic of cause and effect, words like 'goal' or 'aim' raise imponderable obstacles. But engineers and physiologists know that self-regulation without a goal is nonsense – imagine an autopilot on an aircraft that had no idea what height to keep or where to go. .

The Australian physicist Garth W. Paltridge has shown that planetary environments are naturally selected to maximize the planet's production of entropy; in simple words, to keep it tidy and with an orderly balance sheet for energy. Living organisms catalyse the rapid achievement of this goal and at the same time drive the evolution of the whole system. Paltridge has provided another way to approach Gaia theory.

If we are to understand the climate and adapt to its changes, or even counter them, we have to see the Earth as something able to resist adverse change until the going becomes too tough and then, like a living thing, escape rapidly to a safe haven. Fight or flight is a characteristic of life, and the Earth itself, Gaia, has long been resisting our interventions through negative feedback; opposing the way we change the air with greenhouse gases and take away its natural forest cover for farmland. We have been doing this ever since we were hunter-gatherers equipped with fire, but until the last hundred years there was little or no perceptible change in the Earth's state. Now our interventions are too great to resist and the Earth system seems to be giving up its struggle and is preparing to flee to a safer place, a hot state with a stable climate, one that it has visited many times before. A look at the Earth's climate history tells us that in such hot states Gaia can still self-regulate and survive with a diminished biosphere.

It is too often wrongly assumed that life has simply adapted to the material environment, whatever it was at the time; in reality life is much more enterprising. When confronted with an unfavourable environment it can adapt, but if that is not sufficient to achieve stability it can also change the environment. We are doing this now by adding greenhouse gases to the air and by altering the land surface by farming; the outcome is global heating. If the hotter Earth now were more productive than the cool Earth before the Industrial Revolution, we would be flourishing and so would the Earth. Unfortunately we moved the temperature the wrong way, and we may be eliminated as a result. Cooling would have been much better, even though we would have had to abandon much of the northern temperate land to the glaciers. This is how Gaia keeps a habitable planet: species that improve habitability flourish and those that foul the environment are set back or go extinct.

I have slowly come to conclude that scientists are uncomfortable with Gaia theory because it is a threat to the course of their daily lives. Earth scientists, for example, have built for themselves a consistent world where all can be explained by a knowledge of the properties and history of rocks. It coexists comfortably with the life sciences through the use of fossils as tracers and markers of the history of the rocks. By using physics, geologists have discovered the true age of the rocks using radioactive elements as time clocks. If an element like uranium changes through radioactivity into lead at such a rate that half of it has changed in 4.7 billion years, then from the proportions of uranium and lead in a rock we know the date at which it formed. By separating the isotopes of these elements, not a difficult task using a mass spectrometer, the subtlety of such measurements is immensely enhanced. Using chemistry, we can tell when and where gases like oxygen first became abundant in the air and ocean.

The same was true for biologists, happy with a world that Darwin and his successors had described of organisms evolving by natural selection in a static environment. Science can never be certain, but this was as certain as they needed to be. Gaia, like some tyrannical editor, seemed to be asking them to go back and rewrite the text of evolving life – to change it so that the world in which life evolved was no fixed and unchanging world of geology but was as dynamic as the organisms themselves.

In a way the inhabitants of these two major branches of science were expressing the same urge that is making all of us destroy our niche on the present Earth. We all want to continue business as usual. We would like to live out our lives and enjoy our pensioned retirement. Changing a hard-won way of thinking gathered over a lifetime demands a great deal of justification, and I can well understand why biologists do not want to embrace Earth science and have it trample across their cosy niche; neither do geologists wish to herd a host of organisms in their neat and tidy palaces.

Among scientists, only climatologists tolerated Gaia; this may have been because, like physicians, they are at the sharp end of science and constantly accountable to the public. We expect a great deal of our weather forecasters, but they know the world they are asked to predict is chaotic, and so is predictable only to a limited extent. From the earliest days climate scientists have been open-minded and given support: the first invited talk and paper with Gaia in the title was at a meeting of atmospheric scientists, a Gordon Research Conference in New Hampshire in 1970, organized by James Lodge of NCAR. The outstanding atmospheric and climate scientist Bert Bolin, founder of the IPCC, invited the next Gaia paper, this time with my colleague Lynn Margulis. The paper, 'Atmospheric Homeostasis by and for the Biosphere; the Gaia Hypothesis', was published in 1974 in *Tellus*, the Swedish journal of climate science.

The climate scientist Stephen Schneider persuaded the American Geophysical Union to hold two of its prestigious Chapman Conferences with Gaia as the topic, under the titles 'Scientists for Gaia' and 'Scientists Debate Gaia'. I am truly grateful to Steve for his condescension, in the best and not the pejorative sense of that word. The conference in 1988 at San Diego was for me a great ordeal and I felt very much alone. The second in Valencia in 2001 revealed how far Gaian thinking had come and how far it had to go.

Despite the difficulties Gaia theory slowly gained acceptance, and in 2003 the most senior Earth science society, the Geological Society of London, awarded me their Wollaston Medal and in their citation made clear that the award was for Gaia theory; and in 2005 an invitation from the Ecological Society to join their Fellowship finally put the theory in its proper place as one that unified Earth and life sciences. It has taken science a long time to look at Gaia. Why was this? I think the blame lies mainly with scientists in the nineteenth century, who for their own aggrandizement seized and declared independent the territories of physics, chemistry, Earth and life sciences. This conflict over turf still goes on, and new disciplines keep forming. To expect the unifying ideas of Gaia to be welcome in such an environment was foolish, almost as bad as trying to bring peace to a quarrelling married couple – predictably they unite, but in opposition to the intervener. Not surprisingly, the restored union of the sciences brought forth biogeochemistry and Earth system science. And what is wrong with that? Not much, except to ask, would you have read this book if it had the title *The Vanishing Face of Earth System Science*?

Gaia theory has become firmly fixed in the minds of some important American scientists as 1960s mythology and not science at all. Should you think that this statement is an exaggeration, and no more than my disgruntled opinion, consider these recent critical comments. In 2007, a review of *The Revenge of Gaia* by Brian Hayes, in *American Scientist*, had the scornful title, 'Great Balls of Gaia'. Another otherwise favourable reviewer, the physicist Professor Peter Schroeder, began his review in *Physics Today* with, 'The very word Gaia may be sufficient to scare away prospective readers of this book. To me it conveyed a mysterious entity which would be used by New Agers and not by respectable scientists.' He went on to say, 'So my first task is to dispel such illusions and point out that Gaia theory is the product of scientific observation, and like good scientific theories is subject to test and bears predictive fruit.'

Perhaps most telling was a conversation I accidentally overheard in the staff coffee room of the NCAR:

RESEARCHER: I think that we should call our paper, 'The geophysiology of forest ecosystems'.

SENIOR SCIENTIST: You cannot use that word 'geophysiology', it will ruin your reputation as a scientist – it is just closet Gaia.

And so scorn tends to make Gaia theory the science that dare not speak its name. Yet evidence for the theory is already strong and ordinarily in science we would be acting as if it were done and delivered – as in the purchase of a house when the contracts have been signed and one is waiting for the completion date. But with Gaia so much is at stake that we find it hard to accept and make a move. If it is real it demotes us from ownership of the Earth to being one of many animal species. It still allows us to be important and powerful, but the Earth can proceed without us, whereas without photosynthesizers it would probably soon die. At a lesser level, acceptance casts doubt on the way science is divided into a convenient set of disciplines and makes it inexcusable to continue predicting and planning our future on the basis of the reductionist science of past centuries. These issues are far too big to take on in under a decade. I am not asking my scientist colleagues to give up their rational Cartesian way of thinking that has served them so well and immediately become systems scientists. All I do ask is that they take Gaia science seriously.

7

Perceptions of Gaia

A rarely mentioned drawback to science is how frequently we have to take for true things that cannot directly be confirmed by our senses. We are told that everything is made of atoms, but we can never see them with our naked eyes; and, worse, physicists tell us that atoms exist as waves as well as particles and that almost everything is empty space. We just have to believe in our solid flesh. When I look down from space I will be able to see our planet as it is, something real and solid; but as with atoms I can only infer Gaia's existence from indirect evidence.

To illustrate the pitfalls of perception, let me tell you how ten years ago Sandy and I experienced a disturbingly powerful misconception during a walk along the Cornish coast. A rough, stony path led us forward along the edge of cliffs that fell some 400 feet to boulders and bright patches of sand below. This is one of our favourite walks, part of the South West Coast Path around the southern peninsular that points like a finger at the American continent, 3,000 miles across the ocean. As we walked we were immersed in a gentle stream of cool, clean sea air and our ears were full of the sounds of gulls and breaking waves. It was easy to imagine that this was a scene untouched by the artefacts of man. But it did not last. There on a larger patch of sand, ahead and below, was a caravan. On the beach it seemed monstrous, out of place and in fact illegal. Our vision of peace was shattered; if caravans in a place like this became the rule there would be no escape from the noise, the intrusion and the ugliness of urban life. We walked on in indignation, but as if by magic the ugly caravan dissolved in the sunlight to become a patch of sand, dark rock and a pool of seawater. By a trick of light and scene, our two minds had simultaneously perceived the false image of a caravan and our feelings and prejudices had filled in the details, making it seem real. Had we, or the path, turned away from the view, Sandy and I would have stayed convinced that what we saw was real, and would have made certain and confident witnesses in a court of law.

This was a small but unforgettable illusion. As I write now I wonder what separates illusion from what we call reality in our minds. How much of Gaia will I be seeing when I look from the porthole of Richard's spaceship at the Earth some sixty miles below? One way to answer this is to consider how we perceive, and to do that we need to return to the start of our lives. At some time, even perhaps when we are in our mother's womb, our minds follow the instructions laid down by our genes and begin the huge task of building a model of the world based on the constant input from our senses. When we are born we start with more than an empty mind – a great deal of the mind's operating system is specified in our genes and we call it instinct; such as the fear that grips most of us when exposed at the top of a high place above a vertiginous fall. Some species, birds for example, seem to be born with more instinct than we have, and know without being taught how to build their nests, or how to navigate across the world to a distant nesting site. Make no mistake, conscious animals are model-makers: they have to be to survive; and intelligence is mainly a component of our survival kit, as necessary as spines are to hedgehogs, or white fur to a polar bear. I suspect that one huge advantage of human brains is their plasticity, the ability to draw in new information and from it form intuition, a mind-made software that acts as a surrogate for instinct and allows rapid and unconscious action. Unlike instinct, intuition is for the user only and not passed on to the next generation, but it is wonderfully adaptable and it strengthens our mind's model. The adaptability of intuition is especially valuable for a species like ours that wanders across the Earth through ever-changing environments. Perhaps our most important need is a rapid capacity to recognize life. Something alive could be our predator, a tiger well disguised by camouflage and almost invisible against the forest vegetation. It could be our mate coming to a tryst, or it could be our next fresh meal. Our survival and that of our species depends upon a fast and accurate answer to the question: is it alive?

So how do we know, how do we instantly recognize life? Mainly by seeking differences and similarities between what our model predicts and what our eyes see. With a static scene the model in our minds matches the sensory input from our eyes until something in the scene is revealed as different from the background. As every predator knows, movement is revealing. The movement and the shape of the mover indicate the probable presence of life. Rocks, soil and vegetation (apart from when it moves with the wind) are still, and they provide a background of constancy against which the movement of a mammal, bird or reptile is immediately seen and we match its shape against what our model confirms is a match for one of the edible, loveable or lethal parts of life. We rapidly distinguish all plants from rocks and soil in a similar way by their intricate repeating patterns of leaves and stalks: perhaps this is why natural crystals are so fascinating – they are dead but have a repeating regularity not found in mere rocks or stones. The power of our life-detector proves itself when we look into a fast flowing river from a bridge: the constant motion of the water flashes in our eyes as eddies and waves reflect sunlight from the river, yet if the water is clear we can see a fish, especially if it is swimming upstream against the flow, and we know that it is alive. Should you think this easy and obvious and boring, try to design a life-detection device that would detect the presence of that fish. It is far from easy, yet life detection is a free part of our mental equipment and can be updated and improved by training.

We are forever comparing the world perceived by our senses with the world of the model in our minds. When the match is good we accept this as reality. Sandy and I were sure that the caravan we saw on the beach was real, but it was no more than constructive editing by our minds to make a better match to our mental model by touching up a vague part of the scene we thought we saw. It is daunting to think that, had we not walked on and seen that the caravan was a delusion, the next time we walked along those cliffs we would have wondered when and where it had gone. Our model of the world is continually updated, and not always by the truth.

A substantial part of science has arisen from the discovery of instruments that can see, feel and hear far beyond the range of our senses. The microscope that Leeuwenhoek first made in the seventeenth century enabled him to see tiny organisms swimming in a droplet of water, and their shape and motion told him that they were alive. His microscope extended our range of perception to things smaller than we can see by naked eye. Others, especially Galileo, did the same with telescopes, and so now we can see almost to the edge of the universe. Our eyes and ears are limited by evolution to what is needed to survive and nothing more. Human beings have never encountered environments that would have made necessary the possession of eyes that could see in the far infra-red or ultraviolet parts of the spectrum, but reptiles and insects did. We too would have such senses had the need been great enough.

Wondrously powerful as they are, our brains did not evolve by natural selection to see and recognize atoms or distant galaxies. No wonder we scientists try so feverishly to build models and instruments potent enough to make the perception of these imperceptibles seem real. Why then do all of us not see by instinct or intuition something as important as Gaia? Mainly, I think, because until quite recently it has not been important for our selection as organisms. The same is true of the material Earth; it was not until at least the time of the Greeks that wise men told us that we walked on a spherical planet that orbited the sun. Our world previously was no more than the perceivable environment, and what our imaginations made of the sky and the ground beneath our feet.

Friendly scientists often ask me: Why do you keep on talking about the Earth as alive? This is a good question, and there is no rational answer; indeed to some of my friends my suggestion that the whole planet is alive is not only 'scientifically incorrect', it is absurd. In reply I say that science has not yet formulated a full definition of life. Physicists and chemists have one definition, biologists another, and neither is complete. But this does not persuade many of my friends because they think they know by instinct or intuition what is alive, and in no way does the Earth meet their criteria for life. Instinct and intuition are powerful and cannot be denied, and so my assertion of planet-sized life is discounted as an eccentricity.

Well maybe it is, but scientists do not do much better. Thus the physicist Schrödinger, in his remarkable small book *What is Life?*, suggested that a long-sustained dynamic reduction of internal entropy distinguishes life from its inorganic environment; and this thought has been echoed by other physicists, especially Bernal and Denbigh. Biologists simply say a living thing is one that reproduces, and the errors of reproduction are corrected by natural selection. Neither of these definitions is helpful. The physicist's answer is too broad and would mean that mechanical devices such as refrigerators were alive; the biologist's definition is too narrow and would mean that I, a grandmother or a Lombardy poplar tree were dead, since we cannot reproduce. Gaia fits the physicist's definition but fails the biologist's test because it does not reproduce, nor can there be natural selection among planets. But something that lives a quarter the age of the universe surely does not need to reproduce, and perhaps Gaia's natural selection takes place internally as organisms and their environment evolve in a tightly coupled union. Carry that thought further by thinking of a grandmother too old to bear children: according to the biologist's definition she is not alive but she is, as is Gaia, a vast community of cooperating living cells that do reproduce. Perception and insight still set the limit to our wisdom.

Science is broadly divided between the rational Cartesian thinking of Earth and life scientists and the holistic thinking of physiologists, engineers and physicists. The holistic scientists speak in mathematical languages and are all too often incomprehensible to the rationalists. Rational scientists dislike insights; they much prefer step-by-step explanations based on reliable and orderly data. Insight they see as the child of intuition, something irrational drawn from a mess of apparently conflicting data. Dislike it they may, but the large steps in science come as often from insight as from rational analysis and synthesis. This is especially true of quantum physics and of the science of living things; indeed it may never be possible to define life or quantum entanglement in rational scientific terms. Charles Darwin recognized by insight that the evolution of all living organisms is governed by natural selection or, as Jacques Monod put it, through the operation of chance and necessity, but it was not until fifty or more years later, and after a lifetime of research and evidence-gathering by Darwin himself, and later by Mendel, that the full scientific significance of evolution was established by such able men as Fisher, Haldane and Ernst Mayer, and more recently John Maynard Smith, Robert May and Bill Hamilton. Even then it was not until 100 years after Darwin that his apostles E. O. Wilson and Richard Dawkins made it publicly comprehensible. We now have the insight of Gaia that Darwinian evolution is constrained by feedback from the material environment. Thus simply by breathing we add carbon dioxide to the air, which has consequences for everything alive on the Earth, including us, and for the evolution of the whole great system. Gaia is a holistic concept and therefore unpalatable to rational Earth and life scientists. Physicists and physiologists, used to handling the literally incomprehensible, take Gaia and other holistic concepts as useful and are glad to work with them.

My reason for persisting in calling the Earth Gaia and saying it is alive is not a personal foible; it is because I see this as an essential step in the process of public, as well as scientific, understanding. Until we all feel intuitively that the Earth is a living system, and know that we are a part of it, we will fail to react automatically for its and ultimately our own protection. It was not until 2004 that a few of us around the world, including Tim Flannery and Al Gore, came to the insight that climate change was more than an academic scientific project but instead a menacing reality, and one that threatened all of us. Before 2004 the debate about Gaia concerned only me and a relatively small number of scientists, but now a proper understanding of the Earth as a living planet is a matter of life or death for billions of people, and extinction for a whole range of species. Unless we accept the Earth as alive, with us as a part of it, we may not know what to do or where to go as the ocean rises on a hot dry world. For this purpose the name Gaia is far more suitable for a vast live entity than some dull acronym based on rational scientific terms. In ancient Greece, Gaia was the goddess of the Earth. To many Greeks she was the most revered goddess of all, and interestingly the only god or goddess that was never the subject of scandal.

May I remind you why I call the Earth Gaia? It came about in the 1960s when the author William Golding, who subsequently won the Nobel and many other prizes, was a near neighbour and friend.

We both lived in the village of Bowerchalke, twelve miles south-west of Salisbury in southern England. We would often talk on scientific topics on walks around the village or in the village pub, the Bell Inn. In 1968 or 1969, during a walk, I tried out my hypothesis on him; he was receptive because, unlike most literary figures, he had taken physics while at Oxford as an undergraduate and fully understood the science of my argument. He grew enthusiastic and said, 'If you are intending to come out with a large idea like that, I suggest that you give it a proper name: I propose "Gaia".' I was pleased with his suggestion – it was a word, not an acronym, and even then I saw the Earth as in certain ways alive, at least to the extent that it appeared to regulate its own climate and chemistry. Few scientists are familiar with the classics, and are unaware that Gaia is sometimes given the alternate name 'Ge'. Ge, of course, is the prefix of the sciences of geology, geophysics and geochemistry. To Golding, Gaia, the goddess who brought order out of chaos, was the appropriate title for a hypothesis about an Earth system that regulated its climate and chemistry so as to sustain habitability.

My first book, *Gaia: A New Look at Life on Earth*, was written in the 1970s, mostly in Ireland. Perhaps because of the deeply religious sensibility of that land, I wrote: 'There is no set of rules or prescription for living with Gaia, there are only consequences.' This was an insight and not a logically drawn scientific inference, but nothing has happened in the thirty years since to make me change my mind. Our understanding of the Earth has been hampered by the rapidity and success of model-making in computers. I am not for a moment suggesting that computer model-making is not a valuable and enjoyable activity; indeed much of modern science would not have happened without it. The difficulty arises because it is as easy to make computer models of the rational science of the twentieth century as it is to make holistic models like the simple Gaia model known as Daisyworld. Once a large computer model is made and produces a believable result – especially if, when run backwards, it successfully predicts the climate of the previous decades – then its forecasts of the future tend to be accepted as true. This is the state of many of the major climate models now in use by the IPCC. Gaia theory is a holistic, whole-system theory, and as such cannot be modelled using the concepts of the Earth or life sciences separately. Almost all science other than upmarket physics, physiology and engineering is reductionist; in other words, it is about taking something to bits to reveal its ultimately irreducible parts, such as atoms or DNA. Holistic system science is concerned with intact working systems such as the Earth, living organisms and self-regulating artefacts made by engineers. Apart from these dynamic systems, holistic science is still emerging and not yet commonly used in practice.

Computers were first used in science by physicists to help them with difficult equations and with the mind-breaking complexities of the newly emerging concepts of quantum theory. It was not long before engineers with just as hard but more down-to-earth problems used them to improve their inventions, and later they built models that displayed a three-dimensional image of their widgets on the computer screen, images that could be rotated and poked at on the screen almost as if they were real. Engineers are practical people and I doubt if any of their models, no matter how real they appeared, went into mass production without the trial and test of a solid prototype. Other scientists began to compose models and use them to refine their ideas and experiments.

In the 1960s and 1970s computers were hardly more powerful than a pocket calculator, and their program languages were peculiar. One such form of mathematical logic rejoiced in the name 'reverse polish', and not surprisingly non-mathematical scientists avoided it. By the 1980s fairly powerful computers were mass produced and easy to use. Just as the average driver has no idea of the working of a modern car, so scientists using the computer on their desk have no idea of its detailed working but confidently drive it to solve their problems. Earth and life scientists used computers to model the cycles of the chemical elements or the evolution of populations. Computer models are so helpful that before long many biologists and geologists put their field equipment in store and began a new life working with their models pretending that they were the real world. This Pygmalion fate – falling in love with the model – is all too easy, as generations of the young and old playing their computer games have found. Gradually the world of science has evolved to the dangerous point where model-building has precedence over observation and measurement, especially in Earth and life sciences. In certain ways modelling by scientists has become a threat to the foundation on which science has stood: the acceptance that nature is always the final arbiter and that a hypothesis must always be tested by experiment and observation in the real world.

The slowness to accept Gaia theory was also due, I think, to the longevity of the ideas of genius. Just as the elegance of Newtonian physics delayed the emergence of modern physics, so did a rigid interpretation of Darwinism delay the acceptance of Gaia. We have a saying in science: 'The eminence of a scientist is measured by the length of time he holds up progress.' The overarching genius of Descartes, the father of reductionism, still hampers the emergence of a holistic Earth science in which Earth and life science form a single discipline. His insistence on the separation of mind and body persisted as an influence so strongly that only in the last few years has the notion of 'plasticity' become respectable: the concept that thought can change the physical structure of the brain and vice versa.

Physicists and chemists make models but are usually aware of their limitations and almost always ask for experimental verification. Unfortunately Earth and life scientists can only rarely experiment directly with the Earth and are forced to be less pure. Too often the programs that define a model are composed by professional computer scientists or are even commercial modelling applications. The ideas included in the models may be those of the scientists, but the models may be mathematically incapable of handling them. It is as if we expected a car built to travel on roads would do as well across farmers' fields and hedges. Ideally scientists should be personally engaged in the writing of their software, for in this way the model-maker has the chance to interact with the model and maybe even understand it.

Trust in the validity of models made in isolation by Earth and life scientists had a malign effect on their understanding of the Earth. This was because life scientists failed to include a dynamically responsive environment and Earth scientists failed to include organisms that evolved and responded dynamically to environmental change. There was a fundamental and less excusable reason for their reluctance to engage in transdisciplinary modelling. The mathematics of dynamic self-regulating systems frequently involves differential equations that are difficult or impossible to solve by traditional methods. It is too easy to slip into the practice of making what are called 'linearizing approximations', and then forget their presence as the model evolves.

Scientists of these separated disciplines should have realized that they were on the wrong track when quite independently the geophysicist Edward Lorenz, in 1961, and the neo-Darwinist biologist Robert May, in 1973, made the remarkable discovery that deterministic chaos was an inherent part of the computer models they researched. Deterministic chaos is not an oxymoron, however much it may seem like one. Up until Lorenz and May started using computers to solve systems rich in difficult equations almost all science clung to the comforting idea put forward in 1814 by the French mathematician Pierre-Simon Laplace that the universe was deterministic and if the precise location and momentum of every particle in the universe were known, then by using Newton's laws we could reveal the entire course of cosmic events, past, present and future. The first indication that this was too good to be true came in 1890 when Henri Poincaré studied the interaction of three bodies held together by gravity while orbiting in space; he found that the behaviour of the system was wholly unpredictable. This was a serious flaw in the concept of determinism, but it was not until 1961 that Lorenz used an early computer to demonstrate the chaotic behaviour of weather and found it to be wholly unpredictable beyond about a week. He was the originator of the 'butterfly effect' – the idea that the small eddy made by the flapping of a butterfly's wings could initiate much later a hurricane; he showed that this was because weather systems are highly sensitive to the initial conditions of their origin. May found that computer models of population growth showed similar chaotic behaviour, especially in biological systems containing more than two species; these discoveries stirred great interest among mathematicians and scientists in the nature of deterministic chaos. Practical applications in communications and to new art forms have emerged, for example those stunning illustrations of fractal mathematics such as the Mandelbrot set. It was so human and apparently understandable that neither of these eminent scientists made much of the fact that the appearance of chaos suggested that something might be wrong with their hypotheses about the world. Lorenz and May were both looking at the Earth system from within separated scientific disciplines that took cause-and-effect determinism for granted. Yet if instead we look at climate and population growth as a single tightly coupled system we find the combined model is resilient to perturbation and makes credible predictions. This is why I persist with my pleas to IPCC scientists to include in their models the Earth's ecosystems in a similarly tightly coupled, responsive way.

In no way do I imply that either Lorenz or May had blundered into chaos. They were the best of scientists and came upon chaos serendipitously; they had the wisdom to see it as a truly great discovery in itself and one that has enlarged both art and science.

Latest news: Edward and David Wilson's investigation of socio-genesis point the way to the next large step in perceiving Gaia.

A superorganism is something that includes individual organisms but exists as a recognizable entity. It is a category that includes the colony nests of social insects and human cities. The concept of the superorganism could provide a helpful step from the stark individu-alism of the selfish gene to the all inclusive holism of Gaia.

An important part of this step was described by Bert Hölldobler and Edward O. Wilson in their recent book *The Superorganism*. These authors re-define the superorganism as an entity in which the genes of the constituent organisms still rule but evolution proceeds through the selection of the colonies.

The question that interests me is: what sets the bounds of a super-organism like a wild bees' nest? Is it the paper of the nest or the outer layer of bees? As a geophysiologist, I see the whole nest as a live entity, with the paper something the bees have made and that serves to sustain the internal environment of the nest. I would compare it to the shell of a snail or the fur of a bear, something non-living but an integral part of the organism. If it is, then the nest is a form of life, and this is not so preposterous when we consider the observation that bees' nests actively keep their internal temperature comfortable in the cold of winter and the heat of summer.

8

To Be or Not To Be Green

My father, Tom Lovelock, was born on the Berkshire Downs that sit above the small town of Wantage – in King Alfred's time the capital of England. His childhood years were spent almost as close to the natural world as a hunter-gatherer. He talked of a very different world from that described by his contemporary, the novelist Thomas Hardy. Tom was one of a family of thirteen children looked after by my widowed grandmother; my grandfather had worked before his death at a nearby brickyard – owned by the family of the naturalist J. E. Lousley, whose book *The Wildflowers of Chalk and Limestone* graces Collins New Naturalist series. For most of Tom's childhood the family was exceedingly poor. There was no welfare state, and the looming next step down the rungs of poverty was the workhouse. Few now recall the dread of incarceration in the workhouse that blighted the lives of the poor in Victorian times. I have no idea if it was as bad as Dickens portrayed it, but the fear of it was real and held by my grandparents on both sides of the family. My Lovelock aunts and uncles had as children to scrounge a living from the countryside until, much later, my grandmother remarried; my father was one of the older siblings and the main task of gathering food fell upon him. He told me how a local farmer let him take a few turnips from his fields, and what a poor diet they made. He was given the chance to become the apprentice to a poacher and took it; this provided the family with their first meals of meat, as rabbits and pheasants arrived home to supplement their impoverished diet. Few country occupations provide as good training in wild-animal behaviour or, as it is now known, ethology, as poaching. Tom spent much of his childhood years learning this unusual craft. Of course, with no schooling whatever, he was illiterate and innumerate. When he reached the age of fourteen, perhaps made careless by his incipient manhood, he was caught poaching in the local squire's woods by gamekeepers. He was charged with the offence, appeared before the court, and was sentenced to six months hard labour, which he served in Reading Gaol. (This was several years before Oscar Wilde spent time there.)

After this experience he wisely concluded that his trade was now too well known in Wantage, and so he went to London. Here he took a labouring job in the then hi-tech industry of coal gas. The manager of the Vauxhall Gas Works was a chemist, Dr Livsey, and he soon appreciated that Tom was a cut above the average in intelligence. When he discovered that he was illiterate, he sent him to the Battersea Polytechnic, where they remedied his lack of the three Rs. I am proud to possess the handwritten letter from the Polytechnic recording his proficiency. From then on his life rapidly improved in the burgeoning prosperity of Edwardian England.

Tom was the best of fathers, and as the only child of his second marriage I enjoyed his full attention. On weekend walks in the countryside of Surrey, which in the 1920s was only a short tram or train ride from our home in Brixton, I was taught the craft of a countryman by a master ethologist, and from this I developed an intense feeling and appreciation for untouched countryside. This unscheduled teaching must have played a large part in my love of the natural world and in the development of Gaia and all that goes with it.

From Tom I learned the common names of the wild plants, such as lords-and-ladies (*Arum maculatum*), fat hen (*Chenopodium album*), and scarlet pimpernel (*Anagallis arvensis*). Back then on the chalk downs we found *Galium verum* or, as Tom knew it, lady's bedstraw, and sometimes we saw that now almost extinct but splendid scarlet buttercup, the pheasant's eye, *Adonis annua*. Now that we have definition by genome, perhaps those older names will recover the prestige they had before the necessary, but to me dull, nomenclature of Linnaeus was adopted. I think those old names are needed to truly appreciate the importance and seemliness of the natural world, something the punctilious precision of academia has lost for us.

My scientist friends may think me odd to feel this way about proper botany but the poet Ogden Nash put my feelings well in his verse:

> I give you now Professor Twist,
> A conscientious scientist.
> Trustees exclaimed, 'He never bungles!'
> And sent him off to distant jungles.
> Camped on a tropic riverside,
> One day he missed his loving bride.
> She had, the guide informed him later,
> Been eaten by an alligator.
> Professor Twist could not but smile.
> 'You mean,' he said, 'a crocodile.'

Sandy and I now live at Coombe Mill. It was originally no more than two workmen's cottages, which a previous owner had joined together to make a passable three-bedroomed house. It was built on what people in Devon call the hams, the water meadows of the River Carey that flowed past the mill. When I arrived in 1977 the mill and the miller's house were both derelict and beyond any hope of restoration. All of the buildings at the site were constructed of cob, a mixture of mud and straw that Mexicans would recognize as adobe: an efficient but natural building material that keeps the house warm in winter and cool in summer. The climate here is notorious for its lashing rain and high winds, not an ideal climate for a mud-and-straw house, but my friendly local builder told me, 'She'll be fine so long as her head and her feet stay dry,' and despite the wind and rain the cob walls of our cottage have remained strong and supporting for over 250 years. The roof was made with Delabole slates and many of them are still there; these and a few courses of local stone at their feet kept the cob walls dry. The mill house and the mill itself had rapidly deteriorated as soon as the roofing failed to keep the rain out.

Coombe Mill was not the usual grain-grinding mill; it was a workshop with lathes and saw-benches, to which the local farmers brought their timber and had it turned into useful objects. Even though 'renewable' energy – water power – was far less economic than simply buying electricity from the national supply, the mill was a going concern until the early 1960s. Then several things happened.

First the river changed course and the leat that supplied water for the mill wheels dried up, and then the branchline railway that went past the mill closed. The last insult was the arrival of rural electricity, which made pointless any attempt by us to restore the mill.

Before the 1960s Coombe Mill must have been an idyllic place in a natural country way: the mill pond teemed with fish and in the river salmon and sea trout swam and made their nests along the riverbank. Even when we arrived in April 1977 birdsong filled the air, and later in the year the low-pitched buzz of hornets cheered us as they went purposefully on their way and, unlike their cousins the wasps, never pestered or settled on us. We even had otters pay a visit. In many ways Coombe Mill then was the good lifer's dream come true. For Helen, my first wife, then much disabled by multiple sclerosis, the isolation (the nearest house was half a mile away) was a boon after the far too solicitous attention at our previous home in the village of Bowerchalke, 130 miles nearer to London. As a solitary person myself, it was comforting that Helen liked the solitude as much as I did. We were deep in the unspoilt countryside with four-teen acres of pasture, soon to become thirty-five. I was at this time a scientist deeply involved with the depletion of stratospheric ozone by the CFCs. I did not choose to become involved in this politicized environmental concern which was then drawing almost as much attention as climate change does now: the fact was, I had accidentally started it! It happened because I had invented the instrument that measured the abundance of CFCs. As if this were not enough, I had also calculated that these apparently harmless substances were accumulating in the air unchecked. As a consequence it was difficult to avoid working on this atmospheric problem. The greatest pressure to work on the subject came from the odd fact that I was the only scientist in the world who was measuring the atmospheric abundance of the CFCs with any degree of accuracy; this statement may seem false pride but it is not. To carry out my work I needed a laboratory site distant from any accidental releases of these gases, such as from a leaking refrigerator or spray-can. Coombe Mill with its then fourteen acres in a countryside of small and distant farms was ideal. My first task was to arrange with a local architect and builder the con-struction of a laboratory adjacent to the house. By early 1978 this was built and in full operation. But what could I do with fourteen acres of small fields enclosed by ancient hedgerows?

So began my first and last disastrous encounter with biofuels. I had read in *Farmer's Weekly* that Prince Charles had installed on a nearby estate a grass-burning boiler for central heating. Innocently I thought, 'What a splendid way to heat Coombe Mill: with plenty of grass there will be ample fuel, and the surplus could be sold.' I bought one of these specially constructed boilers from a firm in nearby Hatherleigh and had it installed in an outbuilding and plumbed into the house. Cutting the grass and making haystacks of it at the end of the season was done by a friendly nearby farmer – the only problem I had was that he refused payment, saying that he was just helping out a neighbour. The instructions said to put a bale of hay into the boiler, which was a cylindrical tube, and ignite one end as if it were a cigarette. Close the door and it would smoulder and heat the water, and not require replacing for twelve hours. To me this all seemed so economic and so perfectly green: the carbon dioxide emitted from the boiler was that borrowed by the grass from last year's air, so we were merely replacing it.

That was the theory, but in practice I soon found that the boiler was far too much for one busy man to handle because it rarely stayed lit for longer than an hour, and it dawned on me that Prince Charles probably had a retinue of farmhands who would tend his grass-burning boiler and keep it going. In the cold of winter, growing desperate from the demands of this incubus of a boiler, I tried the dangerous expedient of feeding oxygen into the air supply – I knew that even a 1 per cent increase in the oxygen content of the air nearly doubles the chance of the fire continuing to burn. It helped a little but was hardly a green or economic way of central heating. A friendly forester, Mr Thomas, kindly suggested that logs would be easier to handle and that he had plenty. I bought a pile from him at an astonishingly small price, after which life was somewhat easier. Then came a near disaster: in the winter of 1980 the stone path outside the boiler shed was covered in ice, and when I went to deliver a load of logs using my small Iseki tractor it skidded on the ice, fell on to a nearby slope and turned over, pinning me beneath the steering wheel. I switched off the engine (now running upside down) and with a supreme effort managed to drag myself from beneath the wheel. I was in no great pain and could walk around, so I assumed that I had had a lucky escape. The following night in bed I was woken by severe pain in my thigh; foolishly I assumed that my leg muscle had suffered in the accident. Several years later I discovered that in fact I had crushed my left kidney and rendered it dysfunctional.

The tractor was soon restored to health but I still had not worked out what I could do with fourteen acres of grass. Only then did it occur to me that the proper thing to do was to let the land of Coombe Mill go back to nature, to Gaia. Being an impatient man my next mistake was to assume that a return to nature could be hastened by planting trees. I wanted to do it properly, so I sought the advice of a forestry ecologist; he paid a visit to examine the site and sent me a map showing where the different species of trees would naturally be expected to grow – willows and alders along the river, oaks and ash in the fields and a sprinkling of other native trees, including English maple, the wayfaring tree, birch and beech. It all seemed the right and proper green thing to do, and incidentally it would provide a lifetime of green indulgences for travel by air.

Why was it a mistake to plant trees? What I should have done was to have been brave and left the fields untouched so that Gaia in her good time could plant not just trees, but a whole forest ecosystem. A forest is much more than trees: there is the soil with its teeming life at every level from bacteria to worms and beetles, and even moles and badgers; then above ground there are the low-lying plants, the shrubs and of course the birds and animals that occupy what has now become a proper forest. Fortunately, I planted only two-thirds of the now 35 acres with trees – 20,000 in all. On the one-third that was meadow I allowed self-sown trees and shrubs to move in from the edges. There is seemliness about this natural growth that the regimented array of the plantation fails to show. Now the climate is changing, the plantation may not survive, but the natural ecosystem can evolve and change its diversity of species, adapted to whatever is the new climate.

In common with many urban settlers in the countryside, we imagined on first moving to Devon that a four-wheel-drive car would be essential, especially in the icy winters of thirty years ago. We have long since corrected that mistake and now drive a small but roomy Honda Jazz, and travel no more than 6,000 miles a year. The village shop cum post office is two miles away and almost always we walk to it. The weekly shop at the supermarket in Launceston is unavoidable, as are the all-too-frequent trips to Exeter, forty-five miles away, to catch the train to London and the world. Coombe Mill is lit by low-energy fluorescent bulbs and has been for the last thirty years. When not in use our computers are on standby or turned off. I hasten to add, we chose these low-energy devices and actions simply to avoid waste. Having grown up in times of depression and war, the inhabitants of Coombe Mill had developed an instinct towards a frugal life style.

The Lovelocks have been slow in the approach to a true Gaian lifestyle, and the path to virtue is littered with foolish mistakes. Some of these I have listed already, but one I look back on with affection was an attempt in 1978 to lead the 'good life' of subsistence horticulture. At the back of our house is a five-acre meadow and I was stimulated by a magazine article recommending the growing of potatoes under a sheet of black plastic. I placed a 10-metre square sheet of black polyethylene directly on the meadow grass and pegged it down to stop it blowing away in the wind. Next a series of cross-cuts were made in the plastic with a Stanley knife, and seed potatoes were inserted into each of them. The black plastic denied the grass sunlight, so that it died off and became compost for the potatoes to grow in. Grow they did, and in no time healthy-looking leaves and flowers stuck up from the black plastic surface. When the time came to harvest my crop I unpegged the sheet and lifted it from one corner. It really was a splendid crop, and I bent down to lift the first large spud and then leaped back, for the whole sheet was seething with the movement of adders, England's only venomous snake. They had crept under to enjoy the warmth trapped by the black sheet and found an unending source of food in the small rodents who came from round and about to eat the potatoes. Here I thought was the perfect farming ecosystem: grow potatoes this way and have the snakes as guardians of the crop. Sadly it was not to be, for the following winter was by far the coldest we ever experienced. A blizzard with snow as fine as dust blew all one night and cut us off from the rest of England for nearly two weeks. Snowdrifts up to three metres high blocked the roads and the temperature at Coombe Mill fell to −20°C. Next year when summer came the adders and grass snakes were all gone, and so were my hopes of a snake-and-potato farm. I wondered if it was the cold or the heavy snow that had killed them. Whatever it was, it showed that even in the relative warmth of South-West England exceptional weather can still occur. Nothing like it has happened in the thirty years since, but on at least four occasions up until 1990 the fast-flowing River Carey that goes past the mill has frozen hard enough to walk across its fifteen-metre width. Since then the global warming trend has made winters mild with, at most, brief snowfalls that soon melt and frost rarely going below −5°C; the river shows not even a sliver of ice along its banks.

I almost made a similar mistake by imagining that a hornet farm was possible. Soon after we arrived at Coombe Mill we were amazed and slightly scared by the abundance of hornets. English hornets are much bigger than those small, annoying black-and-yellow American hornets: they are huge insects, two inches or more long, with dark brown and reddish-yellow rings around their abdomens. Their sting is potent but (as I have mentioned) their demeanour peaceful. They are carnivorous, eating other insects in preference to jam and sweet food. They are no trouble at all unless you vigorously disturb their nest or sit on one by accident. A local farmer told me, 'You're lucky to have hornets, because if you have hornets you will not be troubled by wasps.' He was right, but like an idiot I then imagined that hornets' nests might be encouraged and the queens harvested to serve as an environmentally friendly way to curb wasps.

Slowly, you see, we had become about as environmentally virtuous as was practical, but realized that it was probably not enough. Perhaps Sandy and I could rebuild the watermill and generate electricity; interestingly our government would now reward us handsomely with a subsidy if we installed a wind turbine, but until recently they discouraged the private use of water power. We have not joined the clamour for green energy and see it as premature, driven by a flawed ideology and the greed of manufacturers and developers who sense easy profits from the subsidies guaranteed by the renewable obligation. We are glad to take the electricity we need from the national grid: home-made energy could be yet another mistake. In our small, densely populated nation the production of electricity at large, efficient power stations is a better option than individual private production. We strongly support a national programme that mainly draws energy from nuclear reactors, by far the least polluting, and now the most economic and reliable of green energy sources. We would welcome professionally constructed nuclear heat at Coombe Mill were it possible for us to have it.

I have discussed and listed the reasons for objecting to the application in the UK of most forms of 'renewable energy' in Chapter 4. But briefly they are because our countryside, before the twentieth century, was a harmonious example of a modestly efficient human niche and was unusually beautiful. The old English countryside was to me the face of Gaia, and something to be sustained in its dynamic form for as long as we were able. Sadly, very little of this now remains, and the land grows every day more urbanized. To make this precious remainder the site of a huge array of industrial-scale wind power stations is to me as philistine as erecting these 400-foot monsters in Hyde Park and the other London parks. As our population grows towards 100 million we will become in effect one gigantic city. Cities need parks and breathing spaces and, looking at South-West England, it would be an act of vandalism to destroy the potential parkland of places such as Dartmoor, Exmoor, the inland small farms and most important the coast of the South-West peninsula to satisfy European green politics. Such a threat is even more absurd when we realize that replacing the two ageing nuclear power plants at Hinckley Point in Somerset would supply securely, cheaply and safely all the energy that the region needs and more.

Sandy and I thought that we were green because we lived the good life in the country and planted trees. We thought that sustainable development and renewable energy sounded sensible. As a young man I had thought, like Alan Bennett, that no intelligent person could be anything but socialist. Now I had moved sideways and my colour had changed from red to green – how could anyone intelligent think otherwise? It was hard to come to terms with the truth that we greens were urban imperialist infiltrators, invading what was left of the old English countryside and with the hubris of true disciples working to change it to our new faith. It dawned on me that we good-lifers were like the Christian missionaries who were the unknowing forerunners of colonial imperialism; like them, we were the vanguard of the urban civilization that soon would conquer the countryside and make the country folk like us. The extent to which I and many of the green movement were wrong was brought home by a short encounter with my farmer neighbour, Billy Daniel, on a walk near Coombe Mill. In a friendly way he said, 'You know you will soon be broke?'

'Why ever do you think that?' I replied.

'No one around here ever made money by planting trees,' said Billy.

He was right. My green good intentions were uneconomic – at least they were then. Ten years or so later the European Community decided to subsidize green behaviour and growing trees became marginally profitable. The Daniels and the other farming families of West Devon were living in fairly stable natural equilibrium as a part of an ecosystem. They had long ago avoided the tragedy of the commons by using hedgerows to establish clear and legal boundaries for their land. Otherwise they were pastoralists, keeping sheep and cattle on the rich grassland.

In the days before Rachel Carson exploded her green bomb, being green was a joyous thing – it was to celebrate the beauty and seemliness of the natural world, and this feeling was surely why many joined or supported environmental movements. Pollution and smog was something that happened to the urban air and made life miserable for those that lived in Los Angeles or London and many other cities. The natural world existed outside the cities and seemed unspoilt and untouched; in no way did we realize that the ever-growing numbers of humans were inadvertently beginning to destroy their world. Then in the 1960s Rachel Carson showed us that the products of our industries – pesticides that farmers used – were massively killing birds in the countryside. *Silent Spring* marked a watershed that separated the old innocent world of naturalists and poets with a sense of wonder at the beauty of the Earth from a new world of ignorant but streetwise urban-dwellers with a sentimental view of nature, who saw destructive intruding species such as the grey squirrel and the Canada goose as needing protection. I would be glad to know if anyone used the word 'green' in the older naturalist sense before it became a word denoting such anthropocentric environmentalism. Before she wrote her book, did Carson think of herself as 'green'? In the 1970s we ceased to be illuminated by the qualities of the natural world and began to see and hear nature through televisual images; often what we saw was filtered or distorted by the thoughts of the presenter. Sometimes we were lucky and saw the real world of nature through the eyes of Sir David Attenborough, but too often it was a politicized account of pollution from industry. Those who were green this way had feelings of guilt and regret; increasing knowledge that once brought wisdom, joy and understanding now confirmed that our carbon footprints were blacker than sin.

From childhood on I have thought of myself as someone who wanted to live naturally and respect wildlife and wilderness. This made me spend much of my free time in the English countryside, and I grew to love it. It was not just the shock of *Silent Spring* that woke me from naivety. The 1960s were the time when mechanization and agribusiness depleted the countryside of wildlife. Farmers have always regarded as undesirable living things that are not livestock, crops, hired help and kinfolk; industry now provided the biocides. Only in the remoter parts of Britain does anything resembling the older countryside remain. For most the delights of urban living began to fill their minds and hasten their departure from nature. Now, in the twenty-first century, the human world of the city dominates and the countryside is merely there as one of its services, to be used as its life-support system: a place for agribusiness farms, sewage disposal plants, reservoirs and now vast alternative energy sites to keep the city illuminated in what seems the perfect green way. What is left of the countryside is fast becoming a set of theme parks with easy access to motorways.

Despite this, I am still a green in that early twentieth-century sense, with views moulded by that older benign philosophy. I know that I am wholly outdated but I acknowledge that I was partly responsible, unconsciously and unintentionally, for the change from a simple green celebration of delight to narrow restrictive faith. When we claim Rachel Carson as the founder of the green movement we forget that like the old school naturalists she had an innocent love of the wilderness and countryside that her several other books reveal, especially *The Sea Around Us*. What caused the emergence of a militant green ideology was the transforming of her warning that our industries threatened wildlife to the message that our industries threatened us as individuals. This is what turned being green from possessing a sense of wonder with concern for the natural world into a partisan and contentious political cause, which at best was no more than a partial expression of the humanism of Christianity or Socialism, and at worst an anarchic extremism. My part in bringing about this change was to enable scientists who used my invention, the ECD, to bring the evidence that unequivocally confirmed the truth of her claim: pesticides and similar poisonous chemicals were everywhere in the world. Later it found that CFCs were accumulating in the air and threatened the integrity of our ozone shield.

Green philosophy has evolved in a complex way. It is still quite diverse and fails to speak with a single, clear voice. The divisions between town and country and wilderness in fact go back at least 2,000 years: Socrates remarked that nothing interesting happens outside the city walls. Generations of natural philosophers, including Ruskin and John Stuart Mill and the poets Blake and Wordsworth, have cursed the consequences of the Industrial Revolution. The early environmentalist lobbies, the World Wildlife Fund, Friends of the Earth and the Sierra Club, were all mainly concerned with wildlife and the loss of countryside or wilderness, and it was not until the 1960s that advances in science made us aware that pesticides and other poisons had spread as far as the penguins of Antarctica, and in fact were in the bodies of everyone in the world. The perceived threat was no longer merely to wildlife; it was now believed to be a real and serious threat to people. It was not long before there was another fusion, between left and green philosophies. The industrial poisons were said to be the products of industries that cared only for profit. The left were then able to say that we were all the victims of those old enemies of Marxism, the capitalists – and now they were not merely exploiting us, they were poisoning us as well. Green intentions were even more distorted when they were combined with those of highly respected anti-nuclear weapons organizations, such as the Campaign for Nuclear Disarmament, CND. Almost everyone agrees that war using nuclear weapons is bad, and this fusion of pacifist and green thinking was also behind the formation of Greenpeace. There were good reasons to object to the mad excess of testing ever larger nuclear weapons, for by 1963 the world had become well and truly contaminated with radioactivity as a consequence.

I was never an enthusiastic supporter of any of these movements because as a working scientist I knew that the sensitivity of the instruments we used to detect chemicals and radiation was so exquisite that traces of them a thousand or even a million times less than a harmful quantity were easily detectable. There was in fact hardly any significant contamination with poison anywhere, except at the bomb test sites or on the farms where pesticides were spread. In Carson's time wildlife, especially birds, suffered greatly from the excesses of agribusiness, not only from pesticide poisoning but also from the removal of hedgerow nesting sites and other massive changes in farming practice that seriously affected their habitats. The green movement was concerned with what happened to people who consumed the farmer's food and only marginally with the natural world.

An unfortunate but inevitable amplifier of the fear of toxic and cancer-causing substances was the way that scientific research was funded in the second half of the twentieth century. Young or ambitious researchers knew that their advancement depended upon the publication of papers that were frequently cited by other scientists and which were interesting enough to titillate the media's appetite for a story. A rich and never-failing seam of pure gold was the discovery that some chemical in common use, such as in a kitchen cleaner, was toxic to an animal species. Then finding this chemical anywhere in the human food chain was enough to start a media scare story, with television and radio interviews and column feet of newsprint. That the amount found in the food chain was often a million times less than a dose that would be poisonous to humans was ignored. There are many different chemicals in every house and so these stories grew; before long they gave employment to lawyers and legislators and became a self-sustaining industry. The inspired storyteller Michael Crichton, in *A State of Fear*, describes this con-junction of vested interests that linked academic scientists, the media, the green lobbies and the legal establishment. In his fiction it is a conspiracy, but in real life the attraction between the parties involved is enough to make conspiracy unnecessary. We must never forget that conjunctions of this kind can do great harm. The unwise and intemperate ban on the use of DDT as an insecticide because of rumours that it might be a human carcinogen is a clear example. DDT sensibly used was so huge a deterrent to the spread of malaria by mosquitoes in the tropics that the ban has been estimated by the World Health Organization (WHO) to have led to the deaths of millions and the disabling of hundreds of millions. Since 2006 the WHO has been actively supporting the use of DDT in combating malaria.

As I have said, in the latter part of the twentieth century the green movement was largely on the political left. Then a slow recognition that capitalism seemed to work better than most forms of socialism, hastened by the collapse of Soviet communism in Russia and Europe, moved green thinking on to a liberal humanist high ground that saw the threat to the Earth mostly in terms of the immediate conse-quences for the welfare of people. There was still no realization that ultimately harm to the Earth system, Gaia, was more serious than harm to humanity. Slowly it dawns that the Earth may be more powerful than we are and now threatens us, and at last we hear the cry, 'We must save the planet!' There is I think a long way still to go.

Nigel Lawson suggests that the present-day green movement is a new religion. To some extent I agree with him, as one who was once an old-style green: I am dismayed by the iconic significance of a giant wind turbine on a green hill. It seems to mock the Cross. You see this new icon in full-page advertisements for the energy companies, especially those companies that sell energy from fossil fuel. It sanc-tifies their hypocrisy and their intention to continue with business as usual; they know that wind energy, because of its inefficiency, is no threat to their main line of business.

In the 1960s I lived for a few years in Houston, Texas, and worked on the new and exciting challenges of space flight. But in my daily life I could not help noticing the all-pervasive and essentially benign ethic of Southern Baptists. Prominent among their beliefs was the harmfulness of alcohol; unlike in Europe, or even New England, the law in Texas then forbade the harmless pleasure of a glass of wine served in a restaurant with a meal. But, amazingly, the law allowed the purchase of a bottle of wine from a nearby liquor store which could be taken to the restaurant to drink with the meal. The liquor industry and the Church had converged on this convenient consensus that allowed the one to profit and the other to stay on its moral high ground. So it is with the convergence of interest between green politics and the energy industry: by spin and small gestures the greens can keep the high meadows and industry prospers. Just as the Houston Baptists failed to save us sinners from the demon drink so the greens are failing to 'save the planet'.

The concept of Gaia or of the world of nature has never appealed to town-dwellers, except as entertainment. We lost contact with the Earth when our food and sustenance was no longer immediately and obviously dependent on the weather. Our fish, meat, fruit and vegetables now come from the supermarket, and only a rare flood or heavy snowstorm impedes a Tesco harvest. When the weather is cold or hot the thermostat takes care to keep our internal environment comfortable. Howling wind and lashing rain against our storm-proof windows can enhance our sense of cosy comfort and not, as they once did, bring fear of a crop lost as the grain was driven into the muddy fields.

Much too slowly some begin to understand that the welfare of Gaia is more important than the welfare of humankind. The science of Gaia confirms the threat to the Earth but allows us to continue the older naturalism where normally the Earth is benign but like ancient goddesses sometimes ruthless, and only humans are sentimental. To be truly green we have to rid ourselves of the illusion that we are separate from Gaia in any way. We are as much a part of her as anything alive and we should feel tied, as in a good and loving marriage, until death us do part.

9

To the Next World

Were he alive today, the American writer Horace Greeley would be exhorting the lively and ambitious to 'Go North, young man'. All the inhabitants of the Americas, including the Indians and Inuits (but not the slaves, who were brought there against their will), are descendants of those who had the courage, fortitude and will to risk the long and hazardous journey to what was then the New World.

Soon people will be migrating again, but this time because their comfort, even their life, is threatened by devastating climate change. We are a wandering species and since our origin in Africa a million or so years ago we have spread across the Earth and even to the Moon. In addition to a natural wanderlust we have been driven to move by seven devastating climate change events as the Earth passed through glacial and interglacial episodes during the past million years. During the last of these, a mere 14,000 years ago, the sea rose 100 metres, enough to inundate a land area as large as the continent of Africa, and the global mean temperature rose about 5°C. William Ruddiman in his books and papers has suggested that the use of fire for forest clearance by early humans may have affected the climate in the distant past. His ideas are plausible and I cannot help wondering if fire-drive hunting in Australia, North America and perhaps East Asia also helped trigger the jump from glacial cold to interglacial warmth. Perhaps the noble savages were just as culpable a cause of climate change as are today's suburbanites.

To communities of fishermen living on the coast of South-East Asia 14,000 years ago, the unstoppable rise of the ocean level must have been heartbreaking when every new settlement was within years faced by eviction. The wiser ones must have moved to higher ground, and some of them were our ancestors. As large a change in our environment is soon due, will be as pitiless, and eventually, in a few hundred years, will lead to a new stable hot climate. As I mentioned earlier, I base this prediction on the Earth's historical record and on models like that illustrated in Figure 3 (p. 33). When it happens the ocean may have risen twenty or even thirty metres, if much of West Antarctica melts into the ocean as well as Greenland; and almost everywhere will be five to six degrees hotter than now. These changes are at least as devastating as was the interglacial shift and will affect a world that is already hot and dry. When they do mass migration is inevitable.

The recognition that we are the agents of planetary change brings a sense of guilt and gives environmentalism a religious significance. So far it is no more than a belief system that has extended the concept of pollution and ecosystem destruction from the local to the planetary scale. Maybe it will grow into a faith but it is still nascent and its dogma not yet properly codified. An environmentalist with a religious inclination might ask, 'Was the discovery and use of fire our original sin? Were we sinful to continue to pollute the planet?' For most of us the contrite expression of 'Mea culpa!' in a deep green voice is not appropriate. We know that we have made appalling mistakes but we have cast aside the old idea that we are born evil and now acknowledge that the whims of our fickle natures were amplified by technology, so that like a drunkard driving a tank we have accidentally trashed our world. Guilt is inappropriate; we seek restitution and the restoration of our lost world, not punishment.

Even if we had time, and we do not, to change our genes to make us act with love and live lightly on the Earth, it would not work. We are what we are because natural selection has made us the toughest predator the world has ever seen. The Tyrannosaurus was displaced even by small mammals. It is as absurd to expect us to change ourselves as it would be to expect crocodiles or sharks to become, through some great act of will, vegetarian. We cannot alter our natures, and as we shall see the bred-in tribalism and nationalism we pretend to deplore is the amplifier that makes us powerful. All that we can do is to try to temper our strength with decency.

Before we discovered and began to use technology, we were the prey of numerous other organisms that kept our numbers in check and so kept Gaia stable by curbing excessive multiplication by any species. But when our ancestors started using fire to cook, they denied a contingent of micro-predators, from viruses to worms, their natural prey. Soon we discovered that the fire kept burning all night scared away the lions and tigers, and finally we began to burn down forests for easy hunting or to clear the ground for farming. Those events hundreds of thousands of years ago mark the time when we broke the rules of Gaia and our numbers grew out of control. We have continued to break them ever since.

When we are first infected by fatal disease organisms, they grow in our bodies without our noticing. We call this the incubation period, and it can be as long a several weeks. Then at some stage in their growth, or in our bodily reaction to it, we feel unwell, with fever and pain. Soon, a matter of hours with the most virulent influenza, homeostasis starts to fail and we collapse and die. This is when physicians speak of massive organ failure. In the whole course of fatal disease there is no tipping point but instead a downslide that starts imperceptibly and then grows ever steeper until we fall.

We became the Earth's infection a long and uncertain time ago when we first used fire and tools purposefully, but it was not until about two hundred years ago that the long incubation period ended and the Industrial Revolution began; then the infection of the Earth became irreversible. Ironically this was the time when Malthus first warned of the danger, and James Hutton and Erasmus Darwin first glimpsed the nature of a living Earth.

The disease that afflicts the Earth is not just climate change – manifest by drought, heat and an ever-rising sea. Added to this there is the changing chemistry of the air and the oceans, and the way the sea grows acidic. Then there is the shortage of food for all consumers of the animal kingdom. As important is the loss of that vital bio-diversity that enables the working of an ecosystem. All these affect the working of the Earth's operating system and are the consequences of too many people. Individuals occasionally suffer a disease called polycythaemia, an overpopulation of red blood cells. By analogy, Gaia's illness could be called polyanthroponemia, where humans overpopulate until they do more harm than good.

There may be nothing we can do to stop the adverse changes now in progress; we cannot reduce our numbers fast enough and there is only a small chance that, using the remedies of Chapter 5, we could reverse climate change. We can talk of carbon footprints, of renewable energy and of sustainable development; we can try to save energy and hold grand conferences around the world, but are these, however well meant, any more than the posturing of tribal animals bravely wielding symbols against the menace of an ineluctable force they do not understand?

Despite this pessimism we are fortunate to live on a planet that keeps itself habitable: the hot state to which Gaia may retreat is uncomfortable but not lethal. We are like the sorcerer's apprentice, unable to undo the industrial spell we cast; yet in time, exposed to the fierce selection pressures soon to come, we, as a species, may grow up and become capable. Our own history through the repeated trauma of the glaciations, and that of the Earth, makes it clear that life has been hard. The impact of rocks 10 kilometres in diameter travelling at twenty times the speed of sound is devastating enough, but worse than these were the vast volcanic eruptions that covered a sizeable part of a continent with molten lava. One of these perhaps led to the great extinction that eliminated most life throughout the long Permian period of geological history. If these physical hazards were not enough there have been other, earlier disasters attributable to the unexpected and at the time unwelcome overgrowth of simple organisms. In Chapter 3 I described how the appearance of oxygen as a dominant gas was a consequence of the evolutionary success of the first photosynthesizers, the cyanobacteria; although ultimately of real benefit, oxygen was at the time the cause of extensive glaciation as the greenhouse gas methane was chemically removed from the air, and it was also toxic to many early life forms. Oxygen on the early Earth was a pollutant and its copious production using the 'green' energy of sunlight was not much different from our own pollution with carbon dioxide from burning fossil fuel; indeed it was the photosynthesizers' oxygen that made combustion possible.

There were almost certainly other hazardous encounters that we have not yet discovered. After every one of these catastrophes Gaia recovered, taking her own time – sometimes as long as millions of years. During these periods of convalescence there was always somewhere on Earth a refuge for living organisms, a place where the climate and the chemistry still favoured life. And so it surely will be when polyanthroponemia resolves. The survivors of the current crisis will find their new home in oases and places still cool and moist enough for life. Gaia may take hundreds of thousands of years before she becomes again the lush planet we once knew.

So how and why does it happen this way? The universe – or at least the universe we live in – seems run by a most inflexible set of rules and we, as curious animals, have tried to list them and use them to explain ourselves, life, the world and all the things we do or make. We are like one who starts to fill in the blank squares of a crossword puzzle: we are delighted to find the answer to a simple clue, and then set back to find it inconsistent with another word that runs down across it. Proud scientists seem to think that soon they will solve the puzzle and know the rules that run the universe with their theory of everything. The most inflexible of these rules so far discovered is what scientists call the second law of thermodynamics. Indeed I find it odd that this is not the first of their laws. This second rule simply does not allow anyone or anything to grow younger or, as they would say, allow water to run uphill by itself.

Perhaps the selfish genes that William Hamilton and Richard Dawkins described, and that are common to all living things, should have been called wilful, because they are forever trying to break this rule. Especially, they would like to live for ever and they have no pity for the life that carries them, only an insistent urge to reproduce. Critics of these neo-Darwinist biologists have called them judgemental by qualifying as selfish so fundamental an entity as a gene. I think that these critics are wrong. Although metaphorical, the word 'selfish' conveys the desire to reproduce which imbues all life. There is no moral imperative that condemns as bad the gene that instructs a living cell to produce deadly poisons like aflatoxin to defeat its rivals, nor was there any moral objection to the use of fire or later to the invention of weapons of mass destruction. The rules simply say that if anything is possible without breaking the second law then, however unlikely, it can happen. Once it does the chances of repetition increase: disinvention is unlikely.

Our inflexible universe is unlikely to be an imposition or limitation to our freedom set by some imaginary intelligent designer; these powerful constraints limit the dynamic wilfulness of genes and make it possible for all life, including our living planet, Gaia, to have a stable existence, one that Gaia has enjoyed for more than a quarter the age of the universe. Utter chaos, in its full physical sense, not freedom, is the consequence of a universe without rules. Some scientists think that there was natural selection among universes and ours is one that succeeded.

Although incredibly long-lived, Gaia is not exempt from the decrees of the second law and now is quite old and soon, in cosmic time, will die. In the days before antibiotics and modern medicine pneumonia was called 'the old man's friend'. It killed off the elderly quickly and without too much pain. The principal hazard for old people is perturbation, whether an infection, like influenza, or an accident, such as a broken hip, whose physiological consequences overwhelm defence. The same is true of elderly living planets – the fatal events are accidents such as impacts of large, fast-moving objects from space, or gigantic volcanic outbursts. An example is the impacting mountain-sized rock that struck the Earth 65 million years ago that is often associated with the demise of the giant reptiles. Catastrophes like these Gaia could survive when young, but if repeated a few hundred million years from now they could be lethal.

The proximate cause of Gaia's ageing is the ineluctable increase in heat from the sun. Our star, like all stars, burns hotter as it ages, and in 500 million years the radiant heat from the sun will be about 6 per cent greater than now. The heat received by the Earth from the sun will have increased from 1.35 to 1.43 kilowatts per square metre; an increase of 81 watts per square metre – no more than the illumination of a single modest light bulb; barely enough to light a room. Not much, you might think, but compare it with the extra heat now coming from all the greenhouse gases we have added to the air; this adds about 1.5 watts of heat to every square metre, a mere one-sixtieth the heat load facing Gaia in 500 million years time. The sun itself will continue to grow more luminous for another 5 billion years before it burns out in a blaze of radiant energy, leaving a planet-sized dense piece of ash; a white dwarf star. Over the next billion years radiant heat will rise slowly and could give ample time for adaptation and further evolution. Already we have a new photosynthetic process due to the evolution of a new class of plants, which biochemists call C_4, able to live at much lower carbon dioxide abundances. In 1982 Michael Whitfield and I calculated that this evolutionary step might enable the contemporary biosphere to continue for another 100 million years. Beyond that, further genetic changes would surely extend the lease of life, but given the fundamental limitations of mainstream biology and the inevitability of perturbations, it is difficult to see life extending beyond 500 million years. Perhaps I am too pessimistic and long before this time has passed some bright intelligence may evolve within Gaia able to keep her alive well beyond that age.

There are organisms called extremophiles which thrive in utterly hostile environments such as boiling water, strong acid or the saturated brine of salt lakes. Optimists among biologists seem to think that a self-regulating Earth system could be based on these organisms when the Earth grows hot. I think that they are wrong because the extremophiles are organisms that have evolved to fill a niche but depend entirely on Gaia to maintain the environment and supply their needs. To expect them to run the planet is somewhat like expecting the proprietors of casinos to run the economy when all other industry has gone. In the same way Gaia can never be based on sparse life. Unless the turnover of chemicals by organisms is as large as or larger than that of a lifeless planet the self-regulation characteristic of Gaia cannot take place.

If we think of Gaia as an old lady still quite vigorous but nowhere near as strong as the young planet that carried our microbial ancestors, it should make us realize more seriously the danger that we are to her continued healthy existence. Our natural misfortune was identified by the great biologist E. O. Wilson. Allegorically, he wrote about that familiar godlike figure, the super-intelligent alien from another galaxy, who was, like Wilson himself, a naturalist. This alien had been observing life on Earth from one of Jupiter's moons. As he prepared to leave on his long journey home he said to a companion, 'How unfortunate that the Earth's first intelligent social animal is a tribal carnivore.'

Nothing that I have read in a long life better explains our agonizing condition – we have the intelligence to begin to expand our minds to understand life, the universe and ourselves; we can communicate and exchange our deep thoughts and keep them outside our minds as a permanent record. We have all this but are quite unable to live with one another or with our living planet. Our inherited urge to be fruitful and multiply, and to ensure that our own tribe rules the Earth, thwarts our best intentions.

Intelligence is no gift from God or the gods; it has evolved by Darwin's rules of selection as the ultimate weapon that lets us rule the world and secure a safe place to raise our children. The Earth is filled with predators, large ones such as polar bears and crocodiles, small ones such as insects or arthropods, and those smaller still such as microbes. To survive the large predators we could have evolved the capacity to run fast like an antelope, or have an unappetizing skin like a porcupine or a turtle, or we could have fought like a bull. Instead we invested all in brains and, as members of the tribe, evolved to become clever enough to outsmart them all.

Individual intelligence alone is not enough, and our amazing achievements come from the additional ability of our brains to communicate and persuade, so that the thoughts of one or a few can persuade the many to lose their identities and act coherently as if they were a single individual. This powerful amplification of the expressed intentions of a tribal leader can always prevail against an incoherent foe or the natural world. This synchronization of will we share with social insects and termites as well as with flocks of birds and fish, and it empowers us far beyond the possibilities of a single isolated intelligence, even one much more able than we are. This may be why some whale species with brains and neuron-counts far greater than ours have never reached the dominance of humans.

This massive amplification of thought and its conversion to action resembles the mechanism of a laser. In this, atoms or molecules, normally bouncing aimlessly around, are lifted to a higher and singular state from which, given the appropriate signal, they can discharge their accumulated energy in a directed beam. Think of how a mob made of otherwise inoffensive people will when lifted by an orator act coherently as if they were a single animal. Almost all of our achievements come from single acts of genius or leadership amplified coherently by the many. This applies not only to great or appalling acts, but to the numerous mundane things that keep society running: the water and electricity services, and road systems.

We are deeply impressed by the power of our weapons, yet they are puny compared with the most powerful weapon of all: creative intelligence. Consider how many of the great and powerful empires have been brought down by ideas alone. Civilizations destroy themselves with ideologies that, like computer viruses, disable their operating systems. Gibbon looked on Christianity as a virus that disabled the mighty Roman Empire. Could Marxism have enfeebled great states in the twentieth century and caused the death of tens of millions? Now we have the urban green ideology, perhaps the most deadly of them all.

E. O. Wilson's alien naturalist must have known that we would be intelligent enough to compose music fit for the heavens and art to match it; create time-proof verse and drama; and be able to see the limits of the universe and start to deconvolute the message of DNA. But did he know that our greatest discoveries could dissolve the civilization that enabled them? Think of internal combustion and how in its many forms it has brought us to our present fix. I wonder if he could have realized that we could evolve into truly social animals, building and living in our city nests and slowly severing contact with Gaia; ultimately in danger of becoming the real and predatory aliens on what had been the planet of our birth.

However irrational it may seem to scientists, we have in all of us a hunger for an ideology or a religion to provide a sense of purpose and wonder when things are good, and reassurance when things are bad. Belief systems provide a programme, which relieves the necessity of thought at times such as the imminent danger of death, when thought is paralysed. Not surprising then that now the dangers of global heating have entered public awareness, environmentalism shows signs of becoming a faith complete with dogma, icons and simple answers to all environmental problems.

I am aware that by bringing in religion I write though a minefield, but I have to do it because the way we think about the Earth is strongly influenced by childhood conditioning and this affects the way we do our science. In Europe we are mostly secular but still strongly influenced by whatever branch of the Christian religion influenced our childhood thoughts. We share a common humanism but differ considerably about our attitudes to the rest of life and to the Earth. I am specifically talking about the different effects of Catholic and Protestant thinking.

As a child my religious education was from the Society of Friends, and as a student I became a Friend and remained one until about 1947, when the agnosticism of science captured me for good. As a student I was also a member of the university's Roman Catholic Society and benefited greatly from the warmth of their friendship and the rigour of their debates on moral theology. I do not think that I was then or now in any way a bigot.

What makes me raise religion is the looming crisis of the Earth, and how near the end of Gaia's most recent interglacial administration is; one that has lasted only 14,000 years. This crisis is the consequence of putting human rights before human obligations to the Earth and all the other life forms we share it with. Protestant philosophy has its bad side but it redeems itself by providing an environment in science for natural philosophy and for holism, whereas the gifts of Catholicism are humanism and Cartesian reductionism. Both approaches are equally needed in science but as impressionable individuals we cannot help being shaped by the teaching of the formative years of childhood so that we tend to favour one and exclude the other.

Perhaps the clearest and most damning example of the harm caused by this separation comes from the distinguished French biologist Jacques Monod, who wrote in his book *Chance and Necessity*,

Certain schools of thought (all more or less consciously or confusedly influenced by Hegel) challenge the value of the analytical approach to systems as complex as living beings. According to these holist schools which, phoenix-like, are reborn in every generation, the analytical attitude (reductionist) is doomed to fail in its attempts to reduce the properties of a very complex organization to the 'sum' of the properties of its parts. It is a very stupid and misguided quarrel which merely testifies to the holists' total lack of understanding of scientific method and the crucial role analysis plays in it. How far could a Martian engineer get if, trying to understand an Earthly computer, he refused on principle to dissect the machine's basic electronic components which execute the operation of propositional algebra?

Maybe these strong words are now less strongly held, but they serve to express what was and still is an important scientific constituency. It needed a powerfully Cartesian world view to have come up with so wrong a first approach to a computer. As any engineer could have explained, dissection – the taking to bits – is the last means of enquiry of a working system. First you interrogate it through its keyboard or by any non-invasive means. If this does not convince you of the limitation of reductionist thinking, imagine that the Martian engineer was an intelligent computer and that it was about to dissect your brain to find out how you did algebra.

Perhaps the gravest error of monotheist religion, including Islam, is to believe that humans are made in God's image. The implication is that we cannot through natural selection improve. To think of us as the perfect model of sentient life is as absurd as to imagine that the first green photosynthesizers to emerge 3.5 billion years ago were also perfect. By evolving and changing they made possible everything that has happened since; if they had stayed as they were there would never have been trees or flowers or animals and ourselves. Nothing in the universe can be perfect and humans have so far to go to approach perfection that surely the future is full of promise.

The Catholic monk Mendel taught us genetics, and the Anglican scientist Charles Darwin, natural selection and we may see both of their ideas in rapid action as this century unfolds and the Earth moves towards its next state. Let's hope that selection chooses from among us those better able to live with Gaia as well as with each other. Are we yet intelligent enough to be a social animal capable of living stably with Gaia and with ourselves now and on the changed Earth that soon will come? As I see it, our hope lies in the chance that we might evolve into a species that can regulate itself and be a beneficial part of Gaia. I wonder if in the great gene pool of all humanity there are the genes that could be selected to meet this goal.

But for now we are what our genes make us, and are not much different from our tribal ancestors who wandered across the continents, often massively destroying wildlife and forests as they went. The most important thing to accept is that there is no going back. If today everyone everywhere softly and silently vanished away it would take at least another 100,000 years before the Earth returned to a semblance of the world that was there before we discovered the use of fire. We have to understand fully that we are still aggressive tribal animals that will fight for land and food. Under pressure, any group of us can be as brutal as any of those we deplore: genocide by tribal mobs is as natural as breathing, however good and kind the individual members of the mob may be.

For too long we have seen the Earth as an infinite resource, or at least an ample resource until technology finds us an equally useful replacement. We are beginning to glimpse the possibility that it may be finite and soon empty, but still we try to make sure that we at least get what we need from the dwindling remainder. In fact, the Earth is neither finite nor infinite but instead always tries to replace itself, as did the forest in which lived our pre-fire ancestors. The forest supplied them with food and raw materials but they had to pay by living in and with it. For us now, Gaia is like the forest. If we think in these terms we see that fossil fuel is renewable energy. Our error is to take more than the Earth renews.

Among the first intelligent humans who enlarged their lives with fires there may have been those who realized that the forests were finite and said so. I suspect the response was, 'Nonsense, they will last over 10,000 years' – and they did. In the same way, how many now are bothered about what might happen in 100 years?

We are strong and adaptable animals and can certainly make a new life on the hotter Earth, but there will be only a fraction of inhabitable land left compared with that available in 1800. If we follow a pure deep green path and go back to a pre-fire existence very few will survive, and if at any time in the new world we recommence carbon-fuel gathering and usage we would be in danger of destroying ourselves and most of non-microbial life. We can use technology but never so much as to disturb planetary regulation. The resilience of Gaia to perturbation would be reduced on a hot Earth, and a rebirth of twentieth-century civilization would then be a major perturbation.

Our first imperative is to survive, but soon we face the appalling question of who we can let aboard the lifeboats? And who must we reject? There will be no ducking this question for before long there will be a great clamour from climate refugees seeking a safe haven in those few parts where the climate is tolerable and food is available. Make no mistake, the lifeboat simile is apt; the same problem has faced the shipwrecked: a lifeboat will sink or become impossible to sail if too laden. The old rules I grew up with were women and children first and the captain goes down with his ship. We will need a set of rules for climate oases.

We are not, as the puritans would have it, some wretched species deep in sin. We could have a great and proud future as the people from whom some future Adam and Eve may evolve, progenitors of a species closer to Gaia and which might serve within her as our brains do in each of us. We would be an important part of what had become in effect an intelligent planet better able to sustain habitability. Social insects such as bees, hornets, ants and termites evolved to form nests – communities far stronger than crowds of individuals – but in so doing they lost personal freedom and became subjects of their queens. Perhaps in a similar way we would lose freedom at the same time that Gaia gained strength. We cannot now know the chances of this happening, how long it will take or what it will be like to be a Gaian subject. The only near certainty is that we will never evolve this way if we allow ourselves, through inaction or an improper response, to be made extinct by global heating. It has been suggested that but for the great extinction 65 million years ago, lizards might now be the dominant intelligent species. In a similar way we could be overtaken by some small animal now present that survives and evolves to fill the niche we had vacated.

Are we clever enough to know who to select? Do we realize that on the hot Earth Gaia's metabolic needs can be met with a mere million or so humans, enough for the recycling of life's constituent elements? Our justification for surviving in greater numbers is that by possessing intelligence we have the potential to evolve to become as beneficial a part of Gaia as were the photosynthesizers and the methanogens; indeed, to make possible an intelligent planet.

I think that we should immediately reject all thoughts of planned selection. There comes to mind in a flash brave people who made the perilous journey across the breadth of Africa, across the Sahara Desert and then built or acquired boats barely strong enough to carry them across the fifty miles of ocean to the Canary Islands. They represent those with the instinct for survival. We exist now because Gaia did the selecting; perhaps we should leave it to her to continue.

Let us look ahead to the time when Gaia is a truly sentient planet through the merging with her of our descendants. We could then look back in wonder at the miraculous evolution of the universe from blazing hot uniformity to a cold mass of simple chemicals, already selected by the cosmos to be the spare parts for life. Then wonder how these chemicals assembled themselves through a series of improbable steps into transient cycles as insubstantial as a house of cards and how the selection and concatenation of these simpler systems led to the emergence of the first living cell. We could wonder why it took so long, nearly 3 billion years, before the cells began to empower themselves as assemblies that were the ancestors of animals and plants. As a planetary intelligence we have already shown Gaia her face from space and let her see how truly beautiful she is compared with her dead siblings Mars and Venus. We could have a future in communion with our living planet to make her strong again and able to counter the disabling impacts that are due.

Thinking this way, how could anyone be a pessimist and imagine that the global heating crisis is the end for us or even Gaia? We will probably both survive and from our descendants could evolve the wiser species that could live even closer in Gaia and perhaps make her the first citizen of our Galaxy.

Sometime later in this century the survivors may reach a small harbour and dismount from their camels. Moored there they may see a small wooden ship scratching its side as it moves with the ocean's gentle swell against the rough harbour wall. A steady, cooler breeze promises a fair start for the next hazardous part of the journey northwards. The captain says nothing as the survivors board his vessel, but he knows that the near-unbearable rigour of the desert has selected them, the strong in mind and body, whose fitness pays the price of the voyage.

Glossary

ALBEDO

This is a measure used by astronomers of the amount of sunlight reflected by a planetary surface. Albedos range from 1 for complete reflection to 0 for complete absorption. The average for the Earth's albedo is 0.33, but clouds and ice can approach 1.0 and the ocean is less than 0.2. Global heating reduces ice, snow and some cloud cover, which leads to lower planetary albedo, a greater absorption of sunlight and even more global heating. The heat absorbed from sunlight is linked to the albedo, but that does not automatically make a dark forest warmer than the light-coloured desert nearby. Most vegetation has a lower albedo than the planetary average but keeps cool by evaporating water from its leaves.

ALGAE

Algae are photosynthetic organisms that use sunlight to make organic matter and oxygen. The ocean plants are almost all algae; some are single cells and others, like kelp, can exist as huge assemblies of cells as long as 60 metres. The first algae on Earth appeared soon after life started over three billion years ago. Their form was bacterial and these microscopic organisms are still abundant: they are found either as free-living organisms or, importantly, as inclusions called chloroplasts within the more complex cells of plants. Algae are unusually influential in the Earth's climate: they remove carbon dioxide from the air and are the source of the gas dimethyl sulphide (DMS), which oxidizes in the air to become the tiny nuclei that seed the droplets of clouds. Fossilized algae are the source of petroleum. Their growth in the surface waters of the sea is sensitively dependent upon its temperature and if this is above 10 to 12°C the physical properties of the ocean prevent the algae from receiving nutrients and they do not flourish. Algal farms may provide a future source of food and fuel.

BIOSPHERE

The Swiss geographer Edward Suess coined the word 'biosphere' in 1875 for the geographical region of the Earth where life is found. In this sense it is a precise and useful term and similar to 'atmosphere' and 'hydrosphere', which respectively define where air and water are on Earth. In the second part of the twentieth century the Russian mineralogist V. Vernadsky expanded the definition of biosphere to include the concept that life is an active participant in geological evolution, encapsulating this notion in the phrase, 'Life is a geological force.' Vernadsky was following a tradition set by Darwin, Huxley, Lotka, Redfield and many others, but unlike them his ideas were mostly anecdotal. Biosphere is now mainly used, in Vernadsky's sense, as an imprecise word that acknowledges the power of life on Earth without surrendering human sovereignty.

CHAOS THEORY

Certainty and confidence in science marked its development in the nineteenth and much of the twentieth centuries, but now it carries on unaware that the determinism that had so long enlivened it is dead. The recognition that science was provisional and could never be certain was always there in the minds of good scientists. The nineteenth-century application of statistics, first in commerce then in science, made probabilistic thinking more intelligible than faith-based certainties. It took the discovery of the utter incomprehensibility of quantum phenomena to force the acceptance of a statistical more than a deterministic world; this was later consummated by the discoveries that came from the availability of affordable computers. These have enabled scientists to explore the world of dynamics – the mathematics of moving, flowing and living systems. The insights from the numerical analysis of fluid dynamics by Edward Lorenz and of population biology by Robert May revealed what is called 'deterministic chaos'. Systems like the weather, the motion of more than two astronomical bodies linked by gravitation, or more than two species in competition, are exceedingly sensitive to the initial conditions of their origin, and they evolve in a wholly unpredictable manner. The study of these systems is a rich and colourful new field of science enlivened by the visual brilliance of the strange images of fractal geometry. It is important to note that efficient dynamic mechanical systems, such as the autopilot of an aircraft, are essentially free of chaotic behaviour, and the same is true of healthy living organisms. Life can opportunistically employ chaos, but it is not a characteristic part of its normal function.

CONSILIENCE

The most distinguished evolutionary biologist E. O. Wilson, when writing on the incompatibility of twentieth-century science and religion, was mindful of the unconscious need in most of us for something transcendental, something more than could come from cold analysis. He disinterred the long-disused but still warm and worthy word 'consilience', and offered it as something to link the thoughts of reductionist scientists with other intelligent humans, especially those with faith. I think he saw it as the name of a concept that would allow these two apparently irreconcilable concerns to evolve, if not together, at least in parallel. His thoughts are wonderfully well expressed in his book, *Consilience*.

EARTH SYSTEM SCIENCE

A discipline that has grown within the Earth science community among those dissatisfied with traditional geology as an intellectual environment for explaining the flood of new knowledge about the Earth. In particular, Earth system scientists dislike the division of Earth and life sciences into the geosphere and the biosphere, and instead prefer to regard the Earth as a single dynamic entity within which the material and living parts are tightly coupled. This concept, together with its conclusion that the Earth self-regulates its climate and chemistry, was publicly stated in the Amsterdam Declaration of 2001. Earth system science arose from Gaia theory but differs from it by refusing to see habitability as the goal for the self-regulation of the Earth's climate and chemistry.

ECOSYSTEM SERVICES

This phrase was introduced by the biologist Paul Ehrlich and his colleagues in 1974 to acknowledge that an ecosystem was more than a place where biologists could study biodiversity. Ehrlich, like Eugene Odum, viewed ecosystems as local regulators of climate, water and chemical resources. 'Ecosystem services' is a valued term when used in this local sense about an ecosystem such as a tropical forest, but is less useful when applied globally because on a planetary scale geophysical and biological forces are strongly coupled.

GAIA HYPOTHESIS

James Lovelock and Lynn Margulis postulated in the early 1970s that life on Earth actively keeps the surface conditions always favourable for whatever is the contemporary ensemble of organisms. When introduced, this hypothesis was contrary to the conventional wisdom that life adapted to planetary conditions as it and they evolved in their separate ways. We now know that the hypothesis as originally stated was wrong because it is not life alone but the whole Earth system that does the regulating. The hypothesis evolved into what is now Gaia theory.

GAIA THEORY

A view of the Earth introduced in the 1980s that sees it as a self-regulating system made up from the totality of organisms, the surface rocks, the ocean and the atmosphere tightly coupled as an evolving system. The theory sees this system as having a goal – the regulation of surface conditions so as always to be as favourable for contemporary life as possible. It is based on observations and theoretical models; it is fruitful and has made eight successful predictions.

GREENHOUSE EFFECT

Most of the sun's radiant energy is in the visible and near infra-red. The air, when free of clouds and dust, is as transparent to this radiation as is the glass of a greenhouse. Surfaces on the Earth, or within the greenhouse, are warmed by sunlight and some of this warmth is transferred to the air in contact with the surfaces. The warm air stays in the greenhouse mainly because the walls and glass roof prevent the restless wind from dissipating it. The Earth is kept warm in a similar but not identical way by the absorption of radiant heat emitted from the warm surface by the gases carbon dioxide, water vapour and methane. These gases present in the air, although transparent to light, are partially opaque to the longer wavelengths emitted by a warm surface. The greenhouse effect has long kept the surface air warm and, in the absence of pollution, is benign: without it the Earth would be 32°C colder and probably incompatible with life.

HYSTERESIS

A forced system can move from one stable state to another, just as a door when pushed can move from open to closed. When the same system fails to respond to forcing in the opposite direction, as when a door is latched, it is said to be in hysteresis. Many natural and engineered systems show hysteresis, as does the climate system of the Earth and the control system of your home heating. When the temperature of a room is below the set point of the thermostat the heat source switches on and heat flows in until the temperature is a degree or so above the set point, when it switches off. There is then a cooling period to a degree or so below the set point and the heat switches on again. This is an example of hysteresis and the climate system responds in a similar way. This is why the reduction of the carbon dioxide content of the air may not be followed immediately by a fall in temperature.

LIFE

Life exists simultaneously but separately in the realms of physics, chemistry and biology and consequently has no decent scientific definition. Physicists might define it as something that exists within bounds, that spontaneously reduces its entropy (disorder) while excreting disorder to the environment. Chemists would say that it is composed of macromolecules coming mainly from the elements carbon, nitrogen, oxygen, hydrogen, and lesser but required proportions of sulphur, phosphorus and iron, together with a suite of trace elements which includes selenium, iodine, cobalt and others. Biochemists and physiologists would see life as always existing within cellular boundaries that hold an aqueous environment with a tightly regulated composition of ionic species, including the elements sodium, potassium, calcium, magnesium and chlorine. Each of the cells carries a complete specification and instruction set written as a code on long linear molecules of deoxy ribonucleic acids (DNA). Biologists would define it as a dynamic state of matter that can replicate itself; the individual components will evolve by natural selection. Life can be observed, dissected and analysed but it is an emergent phenomenon and may never be capable of rational explanation.

POSITIVE AND NEGATIVE FEEDBACK

Self-regulating systems of any kind, from a simple thermostat-controlled cooker to you yourself, always include something that senses any departure from the desired or chosen state, a supply of energy, and the means to apply force that opposes or encourages the deviation. When the car we are driving deviates from our intended path we sense the deviation and with our arms apply enough force to the steering wheel to turn the car's front wheels back on track: this is negative feedback. If by accident the steering mechanism was faulty, so that turning the steering wheel turned the front wheels so as to increase the deviation, that would be positive feedback. This is often a recipe for disaster, but positive feedback can be essential to make a system lively and rapidly responsive. When we talk of vicious circles we have positive feedback in mind, and this is the state the Earth appears to be in now: deviations of the climate are amplified not suppressed, so that greater heat leads to even greater heat.

ROCK WEATHERING

Mountains continuously grow on the surface as the hot, seething, semi-fluid rocks beneath the surface drive the floating plates of rock into collision. On our timescale mountains are permanent features of the landscape, but in Gaian terms they are short-lived and worn away by the weather. Rocks are cracked by frost, abraded by wind-blown sand and, most of all, dissolved away by rain. The dissolution of mountains by rainwater is called by geochemists 'chemical rock weathering'. It happens because the rain contains dissolved carbon dioxide that reacts with the rocks to make water-soluble calcium bicarbonate. This solution is carried by streams and rivers to the ocean. This fundamentally important sink for carbon dioxide was until about 1980 considered by Earth scientists to be purely chemical. We now know that the presence of organisms – from bacteria and algae on the rock faces to trees growing in the soil – makes a three- to tenfold increase in rock weathering and carbon dioxide removal. It is fundamentally important for keeping the Earth cool and as part of Gaia's self-regulation.

SYSTEM

Webster's New Collegiate Dictionary defines a system as 'an assemblage of objects united by some form of regular interaction or interdependence'. Like the solar system, the nervous system or the operating system of your computer, this is the sense in which I use the word 'system' in this book.

Further Reading

1 The Journey in Space and Time

John Gray, *Straw Dogs* (Granta, London, 2002)

John Gray, *Black Mass* (Allen Lane, London, 2007)

John Gribbin, *Hothouse Earth and Gaia* (Bantam Press, London, 1989)

Herman Kahn, William Brown and Leon Martel, *The Next 200 Years: A Scenario for America and the World* (William Morrow, New York, 1976)

Robert Kunzig and Wallace S. Broecker, *Fixing Climate* (Green Profile, London, 2008)

Mary Midgley, *Science and Poetry* (Routledge, London, 2002)

Oliver Morton, *Eating the Sun* (Fourth Estate, London, 2007)

Fred Pearce, *Turning Up the Heat* (The Bodley Head, London, 1989)

Stephen H. Schneider, *Global Warming* (Sierra Club Books, San Francisco, 1989)

Stephen H. Schneider, *The Patient from Hell* (Da Capo Press, Cambridge, Mass., 2005)

2 The Climate Forecast

Robert Charlson, ed., *Earth System Science* (Academic Press, London, 2000)

Sir John Houghton, *Global Warming* (Cambridge University Press, 2004)

Nigel Lawson, *An Appeal to Reason: A Cool Look at Global Warming* (Gerald Duckworth & Co. Ltd., London, 2008)

Kendal McGuffie and Ann Henderson-Sellers, *A Climate Modelling Primer* (Wiley, Chichester, 2005)

Michael E. Mann and Lee R. Kump, *Dire Predictions: Understanding Global Warming* (DK Publishing, Inc., New York, 2008)

Millennium Ecosystem Assessment Report (Island Press, Washington, DC, 2005)

Sir Crispin Tickell, *Climate Change and World Affairs* (Harvard University Press, Cambridge, Mass., 1986)

3 Consequences and Survival

Sir David Attenborough, *Life on Earth* (HarperCollins, London, 1979)

Richard Dawkins, *The Extended Phenotype* (W. H. Freeman, Oxford and San Francisco, 1982)

Brian Fagan, *The Long Summer* (Granta, London, 2005)

Richard Fortey, *The Earth* (Harper Collins, London, 2004)

Al Gore, *An Inconvenient Truth* (Bloomsbury, London, 2006)

Tim Lenton and W. von Bloh, 'Biotic Feedback Extends Lifespan of Biosphere', *Geophysical Research Letters* (2001)

James Lovelock, *The Revenge of Gaia* (Allen Lane/Penguin, London, 2006)

Fred Pearce, *When the Rivers Run Dry* (Transworld, London, 2006)

H.-J. Schellnhuber, *Earth System Analysis* (Springer, Berlin, 1998)

J. Scott Turner, *The Extended Organism* (Harvard University Press, Cambridge, Mass., 2000)

Edward O. Wilson, *The Diversity of Life* (Harvard University Press, Cambridge, Mass., 1992)

4 Energy and Food Sources

Bruce Ames, 'Dietary Carcinogens and Anticarcinogens', *Science,* 221 (1983), pp.1256–64

Bruno Comby, *Environmentalists for Nuclear Energy* (TNR Editions, Paris, 2000)

Gwyneth Cravens, *Power to Save the World: The Truth about Nuclear Energy* (Alfred A. Knopf, New York, 2007)

Jeff Goodell, *Big Coal* (First Mariner Books, New York, 2006)

Michael Laughton, *Power to the People* (ASI (Research) Ltd., London, 2003)

W. J. Nuttall, *Nuclear Renaissance* (Institute of Physics Publishing, London, 2005)

Joel Rayner, *Basic Engineering Thermodynamics* (Longman, Harlow, Essex, 1996)

6 The History of Gaia Theory

Stephan Harding, *Animate Earth: Science, Intuition and Gaia* (Green Books, Totnes, 2006)

Lee R. Kump, James F. Kasting and Robert G. Crane, *The Earth System* (Prentice Hall, New Jersey, 2004)

Lynn Margulis, *The Symbiotic Planet* (Phoenix Press, London 1998)

Lynn Margulis and Dorion Sagan, *Microcosmos* (Summit Books, New York, 1986)

Stephen H. Schneider and Randi Londer, *The Coevolution of Climate and Life* (Sierra Club Books, San Francisco, 1984)

Steven H. Strogatz, *Nonlinear Dynamics and Chaos* (Perseus Books, Cambridge, Mass., 2000)

David Wilkinson, *Fundamental Processes in Ecology: An Earth Systems Approach* (Oxford University Press, 2006)

7 Perceptions of Gaia

John Gribbin, *Deep Simplicity* (Penguin Books, London, 2004)

Bert Hölldobler and Edward O. Wilson, *The Superorganism* (W. W. Norton, New York, 2008)

Anne Primavesi, *Gaia and Climate Change* (Routledge, London, 2009)

Edward O. Wilson, *Consilience* (Little, Brown and Company, London, 1998)

8 To Be or Not To Be Green

Rachel Carson, *Silent Spring* (Houghton Mifflin, Boston, 1962)

Michael Crichton, *State of Fear* (Harper Collins, New York, 2004)

Edward Goldsmith, *The Way* (Shambhala, Boston, 1993)

Richard Mabey, *Country Matters* (Pimlico, London, 2000)

Richard Mabey, *Beechcombings: The Narratives of Trees* (Chatto, London, 2007)

Jonathon Porritt, *Playing Safe: Science and the Environment* (Thames and Hudson, London, 2000)

Jonathon Porritt, *Capitalism as if the World Matters* (Earthscan, London, 2005)

Richard Rogers, *Cities for a Small Planet* (Faber & Faber, London, 1997)

9 To the Next World

Martin Rees, *Our Final Century* (William Heinemann, London, 2003)

Books on Gaia

James Lovelock, *Gaia: A New Look at Life on Earth* (Oxford University Press, 1979)

James Lovelock, *The Ages of Gaia* (W. W. Norton, New York, 1988)

James Lovelock, *Gaia: The Practical Science of Planetary Medicine* (1991), reprinted as *Gaia: Medicine for an Ailing Plane*t (Gaia Books, London, 2005)

James Lovelock, *Homage to Gaia: The Life of an Independent Scientist* (Oxford University Press, 2000)

Index

bold numbers refer to tables, italic
numbers to figures

adaptation 48, 49, 104
aerosol, atmospheric 35–8, 40
agribusiness 9, 86, 144, 146
agriculture, greenhouse gas 47
albedo, reduction of 46, 47, 163
algae 29, 33, 163
   CLAW hypothesis 111, 116
   ocean fertilization 98
Amsterdam Declaration 117, 165
Andreae, Meinrat 36, 94, 111, 116
anti-nuclear propaganda 70–76
Arctic, loss of ice 7, 10–11, 28

Bali, UN Climate Change
   Conference 4, 16, 47
belief, anecdotal 52–3, 73
Betts, Richard 38, 42
Bhopal industrial accident 72–3
biodiversity 115
biofuel crops 12–13
biogeochemistry 31, 121
biologists, and Gaia 119
Bolin, Bert 3, 120
Brand, Stewart 79, 111
Branson, Sir Richard 2
breathing, greenhouse gas emissions
   47

British Antarctic Survey 42
Broecker, Wally, *Fixing Climate* 11,
   97
Brown, Gordon 90
'butterfly effect' 132

C_4 plants 155
Caldeira, Ken 94, 95, 110, 112
Campaign for Nuclear
   Disarmament (CND) 74,
   146
carbon dioxide
   burial 77, 96
   effect on model Earth 34–5
   and energy production **69**
   Eocene increase 101–2
   production by population 47
   reduction of 32, 33
   regulation 108–10, 112
   removal by algae 29, 33, 98
   sequestration 96–9
carbon footprint 18, 48
carbon trading 48, 50
Carson, Rachel 143–5
CFCs 137, 145
chaos, deterministic 132–3, 164
Chapman Conferences 120
char, burial 58, 99–100
Charlson, Robert 15, 36, 38, 94,
   111, 116

Chernobyl nuclear accident 71, 72–3
China, pollution 37
CLAW hypothesis 111, 116
climatologists, and Gaia 120
clouds
    artificial 95–6
    CLAW hypothesis 111, 116
    condensation nuclei 95, 111
    effect on climate 35–8
coal 79, 83
combined heat and power generation 79
Common Agricultural Policy 90
Common Energy Policy 90
computers 130
Connes, Janine 107
Connes, Pierre 107
Cool Earth 97
Coombe Mill 136–43
    ecosystem 139
    grass-burning boiler 138
    horticulture 140
    tree planting 139
countryside, destruction of 9, 144
Cox, Peter 36, 42
Crane, Robert, *The Earth System* 110
Crichton, Michael, *A State of Fear* 147
Crutzen, Paul 94, 95

Daisyworld model 111, 112–14, 115
Dale, Sir Henry 15
Daniel, Billy 143
Darwinism 6, 31, 115, 119, 127–8, 131
Dasgupta, Sir Partha 5
Dawkins, Richard 111, 128, 153
DDT 147

Descartes, René 127, 130, 131, 158–9
deserts, solar thermal energy 66–7
determinism 132–3
Dickinson, Robert 42
dimethyl sulphide 98, 111, 116
disequilibrium 107, 112
dissonance, cognitive 25, 44
Doolittle, Ford 111
drought 10, 54–5
Dyke, James 115

Earth
    ageing 154
    atmosphere 105, 107, 111–12
    catastrophes 52, 152–3, 154
    effect of carbon dioxide 34–5
    hot state 2, 4, 34, 35, 118
    human carrying capacity 56
    as living system 7, 8–9, 47, 62, 165, 166
    surface temperature 39
eco-warriors 21
Ehrlich, Ann 49
Ehrlich, Paul 49
electricity
    dependence on 16, 17, 88–9
    production 65, 68
Electron Capture Detector (ECD) 145
energy 64–86
    and political power 75–6
    renewable 12, 80–85, 142
Eocene, climate 101–2, 104
Erikson, Brent 13
European Union, renewable energy policy 90
evapotranspiration 37, 38
evolution, Darwinian 6, 31, 115, 119, 127–8, 131
extremophiles 155

Farman, Joseph 42
feedback 167–8
  climate models 34, 35, 100–101
  ecosystems 38
Fells, Professor Ian 65
Festiger, Leon 25
fire 149–51
Flannery, Timothy 128
  The Weather Makers 19
flooding 50
food
  production, greenhouse gas 47
  supply 86–91
  synthesized 16, 87, 100
forecasting climate change 23–45
forests
  clearance 97
  evapotranspiration 38
fossil fuels 64, 77–80

Gaia
  naming by William Golding 1,
    106, 128–9
  perception of 126–7
  see also Earth, as living system
Gaia Theory, history 105–22, 166
Gardiner, Brian 42
Garrels, Robert 110
gas, natural 78–9, 83
genes, 'selfish' 153
geochemistry 108–10
geoengineering 92–104
Geological Society of London, 2003
  Wollaston Medal 120
geologists, and Gaia 110, 119
geophysics 32
geophysiology 31, 100–102
global dimming 36, 102
Golding, William 1, 106, 128–9
Goodell, Jeff 80
Gore, Al 4, 15, 128

Gray, John 6
green ideology 12, 142–7
greenhouse condition 33, 101, 166
greenhouse gas 4, 47
Greenpeace 20, 74, 146
Greenspan, Alan 5

Hadley Centre for Climate
  Prediction and Research 36, 38,
  42
Hamilton, William 115, 128, 153
Hansen, James 3, 5, 15
  carbon dioxide reduction 32
  scientific reticence 74
Hardin, Garrett 62
Harvey, Inman 115
Hayes, P. B. 108
Henderson-Sellers, Ann 42
Ho, Mae Wan 106
holistic systems 127, 129–30, 131
Holland, H. D. 108, 112
Hölldobler, Bert 133
hornets 141
hothouse condition 101
Houghton, Sir John 3, 10
humidity, relative 39
hydrocarbons 77–8
hydroelectricity 71
hysteresis 101, 113, 167

India, pollution 37
intelligence 156–7
Intergovernmental Panel on Climate
  Change 3, 4, 7–8
  forecasting 23–6, 27, 28, 29, 40,
  44
isoprene 98

Jet Propulsion Laboratory 1, 13,
  105
Jones, Chris 42

Kahn, Herman 24–5
Kasting, James 108, 110
Keeling, Charles David 6
Keeling, Ralph 6, 14
Koeslag, Johan 115
Kump, Lee 29, 110
Kunzig, Robert, *Fixing Climate* 11, 97
Kyoto Agreement 8

Lackner, Klaus 97
Laplace, Pierre-Simon 132
Lawson, Nigel, *An Appeal to Reason* 51, 147
leaves, temperature 38
Lehmann, Johannes 58, 99
Lenton, Timothy 42, 115
'lifeboat' world 11–12, 22, 56, 161
Liss, Professor Peter 42, 116
Litvinenko, Alexander 75
livestock, greenhouse gas 47
living space 87–91
Lorenz, Edward 132–3
Lovelock, Helen 137
Lovelock Sandy 73, 79, 108, 115, 123–4, 125, 136, 141–3
Lovelock, Tom 134–5

McGuffie, Kendal 42
magnesium carbonate 97
mankind
breathing greenhouse gas 47
importance to Gaia 21
place in Earth system 6
use of fire 149–51
Margulis, Lynn 13, 108, 111
Marine Biological Association 43
Mars, atmosphere 107
Martin, John 98
Maunder minimum 41
May, Robert 128, 132–3

Maynard Smith, John 115, 128
media, anti-nuclear 71–6
methane 79
clathrates 102
micro-organisms 31, 108
Midgley, Mary 106
Millennium Assessment Ecosystem Commission 42
models
climate change 7, 14, 30, 33–5, 40–45, 129
dangers of 4, 6, 14, 26, 129–30, 131–2
Monod, Jacques 127, 158–9

National Centre for Atmospheric Research 42
neo-Darwinism 111, 115, 132, 153
New Age 106, 111
nuclear energy 16–17, 50, 64, 68–76, **83**

oceans
acidification 41, 46, 94, 102
carbon dioxide storage 97–9
fertilization 98
as indicator of global warming 29, 44–5
oil 77–8, **83**
overpopulation 3–4, 9, 49, 77
oxygen 49, 152
concentration 105–6
ozone depletion 42, 95, 137

Pachauri, Dr Rajendra K. 30, 49
Paltridge, Garth W. 118
Parris, Matthew 70–71
Pearce, Fred 106
perception 123–6
of Gaia 126–7
pesticides 143, 144, 145

petroleum 77–8
photosynthesis 38, 49, 99, 152
Pinatubo eruption, effect on climate
    4, 37, 40, 94
Poincaré, Henri 132
pollution
    effect on climate 35–7
    light 3
polonium-210 75
Polovina, Jeffrey 29
Porritt, Jonathon 106
Potsdam Institute for Climate
    Impact Research 42
Prince's Forest Trust 97

radiation, nuclear 70–71
Rahmstorf, Stefan 7, 26, 42
Ramanathan, Professor V. 37
Rapley, Chris 77, 98
rationalism 127
reductionism, Cartesian 127, 130,
    131, 158–9
Rees, Sir Martin, *Our Final Century*
    41
religion 157–9
Rogers, James 79
Rogers, Richard, *Cities for a Small
    Planet* 87
Russell, Bertrand 44

Saunders, Dame Cicely 46
Saunders, Professor Peter 115
Schellnhuber, John 42
Schneider, Stephen 3, 8, 15, 28, 120
Schrödinger, Erwin 127
Schroeder, Professor Peter 121
Schwartzman, D. W. 109
science
    danger of modelling 4, 6, 14, 26,
        129–30, 131–2
    political manipulation 8

wartime innovation 15–16
scientists
    and Gaia 32, 119–20
    reticence 74
sea level rise
    as indicator of global warming 7,
        26–8, 44–5
self-sufficiency 22, 86
sensitivity, climate models 34–5
Severn Estuary, tidal energy 65
Shanklin, Jonathan 42
Shermer, Michael 52–3
smoke 37
solar energy
    thermal 64, 66–8, 83
    voltaic cells 17, 65, 67, 83
Solomon, Susan 14
stars, light pollution 3
Stott, Peter 36, 48
sulphuric acid aerosol 93–5
sun
    evolution of 154–5
    Maunder minimum 41
    see also solar energy
sunshade, orbital 95
superorganism 133
survival 11–12, 52–63

temperature
    forecasting change 4–5, 32, *33*,
        48
    regulation 108–10
Thomas, Lewis, *The Youngest
    Profession* 103
Tickell, Sir Crispin 61
tide energy 65
trees
    artificial 97
    planting 18–19, 97, 139
tribalism 21, 74, 156, 160
Turing, Alan 13

United Kingdom
   2030 scenario 60–62
   effect of global heating 11, 20
   'lifeboat' world 11–12, 22, 56
University of East Anglia 42
uranium 70
urban living 87–9
urbanization 3, 9–10, 142
USA
   attitude to Earth 13–15
   perception of global warming 14

Venus, atmosphere 107
Vernadsky, Vladimir I. 31, 164
Virgin Galactic 2
volcanoes, effect on climate 4, 40, 94
Volk, Tyler 109
voltaic cells 17, 65, 67, 83
Von Bloh, Wernher 42

Walker, James 108–10, 112
Warren, Steven 111, 116
waste, radioactive 69–71
water vapour 38–9
Watson, Andrew 6, 42, 109, 114
wave energy 65
weathering 108–9, 168
Whitmill, Candida 65
Wilson, David 133
Wilson, E. O. 15, 128, 133, 155, 165
wind energy 65, 81–4
wind farms 12, 17, 18, 82, 142
Wood, Lowell 95
World War II 58–60
   scientific innovation 15–16

Zeebe, Richard 110, 112